CLASSIC GEOLOGY

The Gulf of Corinth

CLASSIC GEOLOGY IN EUROPE SERIES

1. *Italian volcanoes* Chris Kilburn & Bill McGuire
2. *Auvergne* Peter Cattermole
3. *Iceland* Thor Thordarson & Armann Hoskuldsson
4. *Canary Islands* Juan Carlos Carracedo & Simon Day
5. *The north of Ireland* Paul Lyle
6. *Leinster* Chris Stillman & George Sevastopulo
7. *Cyprus* Stephen Edwards, Karen Hudson-Edwards, Joe Cann, John Malpas, Costas Xenophontos
8. *The Dalradian of Scotland* Jack Treagus
9. *The Northwest Highlands of Scotland* Con Gillen
10. *The Inner Hebrides of Scotland* Con Gillen
11. *The Gulf of Corinth* Mike Leeder, Julian Andrews, Richard Collier, Rob Gawthorpe, Lisa McNeill, Clive Portman, Peter Rowe
12. *South Wales* Gareth George

OTHER TITLES OF RELATED INTEREST

Confronting catastrophe: new perspectives on natural disasters David Alexander

Geology and landscapes of Scotland Con Gillen

La catastrophe: Mount Pelée and the destruction of Saint-Pierre, Martinique Alwyn Scarth

Minerals of Britain and Ireland A. G. Tindle

Monitoring the Earth: physical geology in action Claudio Vita-Finzi

Planetary geology Claudio Vita-Finzi

Principles of emergency planning and management David Alexander

Volcanoes of Europe Alwyn Scarth & Jean-Claude Tanguy

The Gulf of Corinth

Mike Leeder, Julian Andrews, Clive Portman, Peter Rowe
University of East Anglia

Richard Collier
University of Leeds

Rob Gawthorpe
University of Manchester

Lisa McNeill
University of Southampton

TERRA

First published in 2007 by Terra Publishing

Terra Publishing
PO Box 315, Harpenden, Hertfordshire AL5 2ZD, England
Telephone: +44 (0)1582 762413
Fax: +44 (0)870 055 8105
Website: www.terrapublishing.net

ISBN: 1-903544-23-8 paperback
ISBN-13: 978-1-903544-23-5

17 16 15 14 13 12 11 10 9 8 7
10 9 8 7 6 5 4 3 2 1

British Library Cataloguing-in-Publication data
A CIP record for this book is available at the British Library

Library of Congress Cataloging-in-Publication data are available

Typeset in Palatino and Helvetica
Printed and bound by Biddles Ltd, King's Lynn, England

iv

Λαψνρα ελληνικα: η λεια της Κορινθου
("Greek loot: booty from Corinth")
from
Εισ Ιτακην παρλιαν (a poem, "On an Italian shore")
by C. P. Cafavy, 1925
(translated by E. Keeley & P. Sherrard;
C.P. Cavafy, *Collected poems*. Chatto, 1998)

Contents

Preface

From any viewpoint, the eastern Mediterranean beckons as a superb natural laboratory within which to study the operation of natural processes and hazards in a landscape that cradled the moral and artistic core of European civilization. Thus, we have active tectonics and related earthquakes, instability and failure of steep tectonic slopes, water-supply problems, soil erosion in badland landscapes, the rise and fall of empire and civilization, and the spectre of current and future climate and sea-level change. So, the area is an Earth and environmental scientist's dream. The Gulf of Corinth in central Greece presents a combination of active and inactive normal faulting and uplifted rift sediments, which enable the processes and products of continental extension to be viewed and understood within a Quaternary context of changing glacial and interglacial climates and sea level. Our general objectives are:

- To provide itinerary stops to facilitate first-hand analysis of an active rift in its full environmental context.
- To facilitate the development of skills in field observation, measurement and interpretation.
- To observe active and recently inactive normal faults, appreciating relationships between fault scale, displacements, segmentation and sedimentation patterns.
- To employ the sedimentary record to assess the evolution of palaeo-environments, sea-level change and climate during the Quaternary.
- To encourage Earth and environmental scientists of all persuasions to visit the area for training and research into the complexity of the Earth system, in particular the use of sites for comparison with more stable crustal settings, with geologically more ancient rifts and for the better exploitation of reserves of natural resources in similar rift basins worldwide.

Our geological booty set out in this guide has poetic licence, hence the central line of C. P. Cafavy´s poem recreating loss after the vengeful Roman republic´s army under Mummius totally destroyed fabled ancient Corinth in 146 BC – the ruins you see today (front cover) are almost entirely Roman, unfortunately.

Mike Leeder, Julian Andrews, Richard Collier, Rob Gawthorpe,
Lisa McNeill, Clive Portman, Peter Rowe

Acknowledgements

Our continuing work in the Gulf of Corinth rift has spawned over 30 papers over the past 24 years; none of this would have been possible without the necessary logistic and sampling permissions from IGME, the Greek Geological Survey. Our onshore geological studies have benefited from perspectives offered by cruises in the gulf with colleagues George Ferentinos, George Papatheodorou, Aris Stefatos and many past and present students from the Department of Marine Geology, University of Patras. Our shipboard experiences include sharing *al fresco* lunches, often cooked in a super-hot galley in between poring over the latest geophysical printouts. Thanks are also due to the Greek National Centre for Marine Research (NCMR) for access to data and help with logistics.

As always, we owe a heavy debt to our present and former undergraduate and postgraduate students and postdoctoral researchers at the universities of East Anglia, Leeds, Manchester and Southampton, and to other students over the years, often crystallizing new ideas, after fieldwork, in those early-evening discussions over ouzo back at hotel or campsite base. So, special thanks are due to Peter Bentham, Alex Brasier, Carol Cotterill, Chris Dart, Pierre Eliet, Karen Knapton, Evrikos Lyberis, Jenny Mason, Leslie McMurray, Dave Ord, Sarah Prossor, Gerald Roberts, Mark Seger, Jenni Smith, Colin Stark, Clare Stephens and Mark Trout.

Establishing an accurate chronology for assessing Quaternary rift evolution was an early major obsession with us, for without chronology it is impossible to understand tectono-stratigraphical evolution properly and to put it into the framework of Quaternary sea-level and climate change. It was following Tim Atkinson´s suggestion in 1987 that we went for U-series disequilibrium dating of corals and we gratefully acknowledge Tim´s influence on and support for our work.

We thank many other colleagues and friends who have helped and collaborated with us, including Emma Finch, James Jackson, Dora Katsonopolou, Ioannis Koukouvelas, Greg Mack, Paolo De Martini, Dan McKenzie, Daniela Pantosti, Marta Perez-Arlucea and Nicky White for discussions over the years. Many delegates from BP, Mobil and many other multinational oil companies have helped us along with research grants and PhD studentships over the years; we especially thank Richard Hodgkinson (ex-Mobil) Mike Bowman and John Dixon (both ex-BP).

Although we have made every effort to ensure that information on localities mentioned in this guide is correct, for the benefit of future editions we welcome from readers any corrections or updates concerning access, state of outcrop and directions.

Chapter 1

General introduction

The Aegean sea, with its bounding tectonic arc, is a diverse natural laboratory for the study of the action of tectonic processes and their effects upon the landscape. Many lessons may be learned here concerning the evolution and interrelationship of faulting, geomorphology and sedimentary deposition, for application elsewhere and in older parts of the geological record. Within the Aegean, the Gulf of Corinth is one of the world's key localities for understanding the interrelationships between continental seismicity and crustal deformation, kinematics of normal faulting, geomorphology, sedimentology, and sequence stratigraphy. The region is ideal for scientific research and training because of the excellent exposures afforded by uplift and incision, not least by the excavation of the Corinth Canal. At the same time one can enjoy the scenery of rugged mountains, picturesque gorges, small coastal resorts and Peloponnisian villages. But the menace and imprint of the Earth's forces is never far away. There were some destructive earthquakes during the twentieth century, including the April 1928 Corinth earthquake, which led to the new town being rebuilt on its present coastal site, the Perachora–Pisia earthquakes of 1981 and, more recently, the Egion event of 15 June 1995 and, farther east, Athens, 7 September 1999, the latter with heavy loss of life in the western suburbs of the capital city.

Places, myth and history

Observations have been made of the changing landscape in central Greece for over 2500 years. Xenophanes (*c.* 570–475 BC) recorded fossil fish and shells on mountains far inland, inferring that either the land had risen or that sea level had fallen; we are still struggling with that one today. The countryside is littered with historical sites and artefacts, some dating back to the Mycenean period (about 1000 BC or earlier). We will pass by some of these and linger on a couple that have rich geoscience associations.

Megara was founded in the Mycenean period and was once the site of

1

the shrine of Hera, Zeus´s consort. The shrine was subsequently moved to Corinth and finally to Perachora. The Megarans were part of the Athenian League; long before that they controlled the strategic land route for pilgrims from Athens, who embarked from the port of Pagae (modern Alepochori) to the famous oracle and shrine of Apollo at Delphi. Megara was also a place of exile for offending philosophers from the city state of Athens.

The modern highway from Megara to Corinth follows the path trod by Theseus, who, according to legend, killed the giant wild Megaran sow, Phaia, and set out to free the bandit-ridden coast road that led from Troizen to Athens. Following the coast road, Theseus came to precipitous cliffs rising sheer from the sea (north of modern Kineta), which had become a stronghold of the bandit Skiron. The cliffs of Skiron rise close to the Molurian Rocks, and over them ran Skiron's footpath, made by him when he commanded the armies of Megara. Graves (1992) notes that a violent northwesterly breeze, which blows seawards across these heights, is still called Skiron by Athenians. Skiron used to seat himself on a rock and force passing travellers to wash his feet; when they stooped to the task, he would kick them over the cliff into the sea, where a giant turtle swam around waiting to devour them. Theseus, refusing to wash Skiron's feet, lifted him from the rock and threw him in the sea. Further airborne events occurred in 1944 when the airfield was liberated from the Nazis by the British army's 2nd Battalion, Parachute Regiment, in a difficult drop during which a VC was won. The paras suffered many casualties, partly because of a brisk Skironian wind.

Perachora was an important religious sanctuary dedicated to Hera. The foundations of the eighth-century BC temple are well preserved, as are the foundations of a smaller sixth-century BC temple, a colonnade with a narrow roof, a small marketplace and some impressive water cisterns in the upper excavations. The incomparably beautiful site (Ch. 4) was excavated by the British School of Archaeology in Athens in the 1930s. The atmosphere of the general area during her late husband´s excavations is movingly related by the erstwhile film critic of the *Sunday Times*, Dilys Powell (Powell 1957, 1967).

Ancient Corinth has suffered destruction, rebuilding and resiting on several occasions. The Roman general Mummius destroyed the Greek city after the Peloponnisian revolt against Roman rule in 146 BC (little survives from before this time). Roman Corinth was sacked by the Herulians in AD 267, the Visigoths in AD 395, and was struck by earthquakes repeatedly (most recently 1858 and 1928). The Doric Temple of Apollo, with seven of its original 38 exterior columns, still stands (cover photo), and the Fountain of Peirene dates back to the Archaic period. The founder king of Corinth was Sisyphus, to whom the river god, Asopus, gave the Peirene spring in return for information concerning the whereabouts of his daughter Aegina, who had been abducted by Zeus. Sisyphus was father of Odysseus and grandfather to Bellerophon, who, riding the fabled winged horse Pegasus, slew the Chimaera. Jason and Medea lived here with their young children, until Jason proposed a marriage of convenience to Corinthian King Solon's daughter Glauka (Creusa). Medea then poisoned Glauka (she tried to wash off the skin poison in what is now known as the Fountain of Glauka) and murdered the children, fleeing to Athens for protection, where she later tried to entrap Theseus. Acrocorinth, the mountain immediately to the southwest of ancient Corinth, formed the acropolis in classical times, being the site of an epic brothel with over a thousand sacred whores. The impressive fortifications seen today date from the Middle Ages, the site passing back and forth between Frankish, Venetian and then, finally, Ottoman colonialists.

Kenchreai, together with Lechaion on the Gulf of Corinth coast, was one of the twin ports of Corinth. Damaged by two earthquakes in the late fourth century AD, it is now mostly submerged beneath the sea. Remains can be seen of an important early Christian basilica. Some say that St Paul read his letters to the Corinthians here during his stay in AD 51–52, chastising them for their wanton ways, but also speaking memorable prose ("O death, where is thy sting" . . . etc.). Loutro Helene, a brackish spring south of Kechriae, was thought in antiquity to be the spot where Helen of Troy once bathed. Copious flow from the spring can still be seen.

Eliki (or Helice), an ancient Greek city on the north Peloponnisos coast a few kilometres east of Egion, was destroyed by an earthquake in 373 BC when " . . . all the mice and martens and snakes and centipedes and beetles and every other creature of that kind in the town left in a body . . . And the people of Helice seeing this happening were filled with amazement, but were unable to guess the reason. But after the aforesaid creatures had departed, an earthquake occurred in the night; the town collapsed; an immense wave poured over it, and Helice disappeared . . . " (*Aelian* 11.19, Loeb trans.). The submerged ruins of Helice were visible for 500 years, but have since been lost, some of them perhaps reworked because of slope failure and dumped in the gulf, although there is no direct evidence for this. The Eliki fault escarpment was reactivated in an 1861, and is thought to be the same fault whose deformation destroyed the ancient city.

Safety in the field

General points:
- Follow all safety directions and make hazard assessments at all localities. A member of your party should have a first-aid kit.
- Each individual in a party must take responsibility for their own safety. If in doubt – for example, over the stability of a cliff outcrop or your ability to stay on a narrow path – then keep back.

Particular concerns:
- Be careful when working on road sections at all times. Wear high-visibility jackets. The roads you may stand on looking at geological features can be narrow and windy. Greek drivers can be very fast. When Greek drivers flash headlights at you, they are *not* saying "After you, please", but "Get out of my way".
- Beware of unstable outcrops. When moisture evaporates in the early morning, pebbles are particularly prone to loosening and failing. Avoid all cliff-like outcrops during rainstorms or high winds. Safety helmets should be used adjacent to cliffs and steep slopes.
- If you are afraid of heights, keep away from cliff edges or the sides of the Corinth Canal and elsewhere.
- Wildlife: When swimming in areas with rock outcrops or large pebbles, take extreme care not to step on sea urchins – their spines penetrate deeply, are difficult and painful to extract, and are likely to go septic. Beware of small scorpions when turning over or picking up rock specimens. You should be making plenty of noise underfoot to scare off any snakes (these are very rare), and beware of the wild tortoise.
- Have plenty of bottled drinking water with you, 3 L per person per day in summer.
- In the fierce Aegean sunshine, grease up and cover up. Wear a hat with a broad brim.
- If you smoke in the field, take extreme care over the fire risk created by dry ground conditions (innocent geologists have been temporarily arrested in the field in Greece on suspicion of arson after development of forest fires in their mapping areas).
- Update your tetanus shot, just in case.
- Emergency numbers: police 100, fire 199, ambulance 166, tourist police 01-171.

Finally, a disclaimer; the authors cannot be held responsible for the safety of any visitors to any of the sites or locations featured in this guide.

Travel and accommodation

The ease and cost of travel to Greece from northern Europe have changed markedly for the better in the past few years with the advent of scheduled budget flights into the new Venizelos airport (Venizelos led the ill fated Greek delegation to the Treaty of Versailles), which serves Athens via a dedicated metro link. Travel to the Corinth rift is now easily achieved by private car on the new toll highway that blessedly avoids Athens city centre, with its attendant distractions and diversions. We have found the cheapest car hire to be courtesy of ArgusCarHire.com,* whose staff are always helpful.

Regarding accommodation, there are plenty of hotels to choose from in most locations. Our favourite in Loutraki resort, a convenient base for localities in the eastern gulf, is the Hotel Sega (tel.: +30 27440 22623; fax: +30 27440 67610; seagasloutraki@yahoo.com) for its friendliness, convenience and outstanding value. The Niko's taverna in ancient Corinth (tel.: +30 27410 31361) not only provides delicious food but there are also comfortable flats to rent there at reasonable prices. The Five Brothers taverna on the northern limit of the Loutraki seafront is an evening haven throughout the year; it features a charming mural of Aphrodite over the kitchen entrance. In the western gulf, Hotel Chris Paul in Diakofto (info@chrispaul-hotel.gr; tel. +30 26910 41715, fax +30 26910 42128) is highly recommended.

Sampling, permissions, map coverage

Following modern codes of conduct for field parties, it is strongly advised that rock samples from solid outcrops be taken only when *strictly* necessary. Greece has specific rules for the conduct of visiting geologists, because geological sites are commonly adjacent to or within world-class archaeological sites and also because local inhabitants often confuse legitimate field sampling with artefact robbing. Thus, field parties intending to sample, and individuals visiting Greece either casually or to do research, need to obtain a permission order (one per party) from IGME (Institute of Geology and Mineral Exploration), which is best done in advance by fax nowadays, and which they should produce when required by any local functionary. We have always found the staff at IGME to be utterly helpful and obliging. Geological maps are also obtainable here (N.B. the map department is closed in the afternoon), but note that these are rather dated

---

* http://www.argusrentals.com

with respect to Quaternary stratigraphy and geological structure. There are strict rules operating within archaeological sites, where sampling and the carrying of hammers and the like may be strictly forbidden, and you should check with the local authorities at the site before attempting any such activity. Research geologists wishing to take multiple samples out of Greece for further laboratory analysis must take them to the IGME offices,[*] where they are inspected for non-inclusion of artefacts and then sealed in an official bag to take through airport check-in.

Obtaining suitable topographical maps for field investigations in Greece is inconvenient compared to their widespread availability in other EU countries. Maps at 1:50 000 are somewhat out of date and not really suitable for accurate mapping. Those at 1:5000 scale are particularly good and detailed, although are often dated with respect to the wave of building and roadmaking activity over the past 15 years. GPS-aided mapping mostly overcomes deficits in the accuracy of these; sometimes we have found deviations up to 50 m with respect to absolute coordinates of latitude and longitude.[†]

---

[*] Contact details: 70 Messoghion St, Athens 115 27 (nearest metro stop Katehaki); tel. +30 210 7798 833,fax +30 210 77 96 586, http://www.igme.gr/enmain.htm. Current contact inthe permits department: Eleni Kotronia. IGME will resite near the Olympic Village in 2007.

[†] All the above maps may be purchased only from the Hellenic Military Geographic Services. Contact details: Evelpidon 4, Pedionareos 1/362, Athens; tel. +30 210 8206633, fax +30 210 8817376, gys@hol.gr, http://www.gys.gr/english/EN4.htm. Opening hours are Monday, Wednesday, Friday 0800–1230. A passport is often required for entry and payment (20 euro per map) is by cash only. The nearest metro stop is Victoria.

Chapter 2

Geological background

Legacy of the Tethys Ocean

The older basement rocks of the study area in central Greece originally accumulated as continental shelf limestones and related sediments during the Mesozoic development of the Tethys Ocean, which once widely separated Africa from Eurasia (Fig. 2.1). The progressive collision of the African and Eurasian plates during the closure of the Tethys to form the Alpine–Himalayan range was not simple, and involved strike-slip, compressional and transpressional motions. The contact of the irregular continental margins involved is an additional complicating factor and several microplates were also involved, including one, the Adriatic microplate (Adria), that included much of southern and western Greece. Structural relations within the Greek Alpine belt are thus tortuous and it has been a challenge to disentangle over the past 30 years the timing and sequence of plate collisions and related deformation.

Tethyan shelf sediments of the External Hellenide and Pelagonian zones (Fig. 2.2) mainly accumulated on marine carbonate platforms of the Adria microplate; they now comprise limestones and marbles with cherts. Serpentinized ultrabasic rocks within ophiolitic bodies occur along major thrust faults in many areas and are of particular importance in the eastern Gulf; they are remnants of the Tethyan oceanic crust and mantle, now marking terrane boundaries within the former Tethys. Thrusting to emplace these occurred during a major mountain-building episode, when crustal shortening and partial closure of a failed-rift branch of the Tethys, called NeoTethys, occurred during late Cretaceous to early Tertiary times (60–30 million years ago), causing formation of the Hellenides, part of the Alpine–Himalayan belt. The thrusting was accompanied by formation of thrust basins, whose infill included marine turbiditic clastic sediments (flysch) and younger terrestrial fluviatile conglomerates. The structural grain (strike) of the Hellenides is generally north–south and the similarly orientated remnant of thickened (40 km) continental crust throughout west-central Greece attests to the substantial mountain range that existed

7

over the whole area of the Aegean prior to the onset of lithospheric extension.

The Aegean arc

Following formation of the Hellenides, northward subduction of African oceanic plate caused the formation of an Aegean magmatic arc, which was responsible for the production of magmas and associated plutonic (granites, granodiorites) and volcanic edifices caused by partial melting above the water-rich subducting slab between 15 and 5 million years ago (mid-Miocene to early Pliocene). As the slab progressively rolled back, rapid

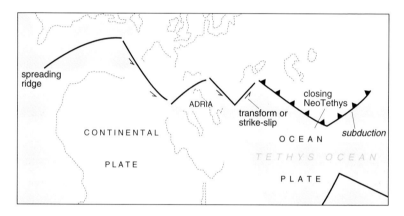

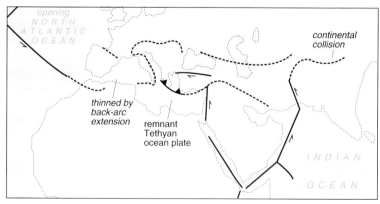

Figure 2.1 Highly generalized sketches to show (upper) the position of the Adria microplate with respect to the closing Tethys Ocean during the early Cretaceous and (lower) the current plate-tectonic setting with remnant Tethyan ocean plate subducting along the Hellenic subduction zone.

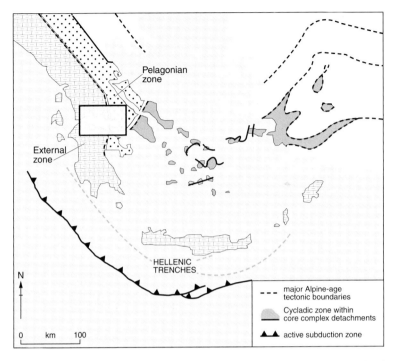

Figure 2.2 Generalized tectonic sketch map to show the context of the Gulf of Corinth (box) within the wider Aegean area, chiefly the major thrust-bounded Pelagonian and External zones of the pre-rift Tethyan and pre-Tethyan basement of Adria and the late-Tertiary core complex detachments of the Cycladic zone (latter after Kumerics et al. 2005).

back-arc extension of the thickened, hot Hellenide crust occurred in all areas north of Crete. Spectacular evidence for this episode of pervasive crustal thinning comes from the existence of core complexes at many localities across the Aegean. Such terranes were first discovered in the western Basin and Range province of the Rocky Mountains in the late 1970s and represent exposures of metamorphic and igneous rocks that once formed part of the mid-crust and which have been exposed by rapid exhumation along low-angle normal faults and shear zones, termed detachments (Fig. 2.2). The process in the Aegean may have led to more than 200 km of extension. Accurate radiometric and thermometric dating suggests this exhumation as having occurred mainly in the interval 10–3 million years ago (i.e. upper Miocene to lower Pliocene).

As extending Aegean crust thinned and cooled, the regime of detachment faulting was replaced between 3 and 1 million years ago by high-angle normal faulting. The extension rate remains high to this day as the Anatolian plate moves rapidly southwestwards, converging at the Hellenic

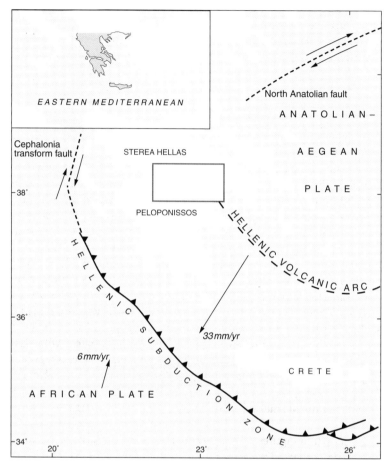

Figure 2.3 The position of the Gulf of Corinth (boxed) within the current plate configuration of the Aegean area. Plate velocity vectors after McCluskey et al. (2000).

subduction zone with the African oceanic slab at a combined velocity of almost 40 mm a year (Fig. 2.3). The time trend from low-angle detachment to high-angle normal faulting is also seen in the western USA, although there the rate of active extension is an order of magnitude less than in the Aegean. The continuance of rapid extension to the present day involves the whole greater Aegean plate, from Crete to Turkey and also northwards into the south Eurasian terranes of Romania. As we shall see, much internal strain is produced during this extension, the majority of which occurs across the Gulf of Corinth rift. During the past 3 million years the Aegean extensional province has been broken up into a series of tilted blocks by

10

major normal faults; these cause block rotations about both vertical and horizontal axes. Recent satellite geodetic studies have much improved previous seismological and geological evidence for the direction and magnitude of the horizontal and rotational motions in central Greece.

General morphology and structure

As we have noted, land areas surrounding the Gulf of Corinth comprise rocks of the External Hellenide and Pelagonian tectonic zones (Figs 2.2, 2.4), the thrust contact between the two being marked by major ophiolitic basement slices, clearly exposed in the Gerania Mountains of our study area. The gulf itself is the major geomorphological feature of central Greece and is the geographical demarcation between Sterea Hellas to the north and the Peloponnisos. The gulf is 120 km long and up to 27 km wide, with a maximum depth of 869 m. The east-southeast–west-northwest trending structure has a broad symmetrical trough below 400 m water depth and a broader northern shelf shallower than this over much of its central area (Stefatos et al. 2002); this northern shelf narrows in the extreme west. Bathymetric gradients down to 600 m are usually steepest along the southern gulf margin, although this varies because of the presence of active normal faults within the gulf and of many base-of-slope submarine fans. East of the connecting symmetrical trough of the Strava graben (Papatheodorou & Ferentinos 1993), the bathymetry of the shallower Alkyonides Gulf (maximum depth 360 m) is strongly asymmetrical because of southerly tilt into major north-dipping normal faults along the southern margin.

As shown in Figures 2.4–2.6, widespread areas of Quaternary (Pleistocene–Holocene) sediments occur in abandoned fault-bounded basins on the southern side of the modern Gulf of Corinth. Both inactive and active basins cut across the north–south structural grain of the older Hellenide mountains. The initial (?early Pleistocene) Corinth Basin floor may have been in an intermontane setting as a lake or series of lakes, at some as-yet undetermined height above contemporary sea level.

The neotectonic structure of the Gulf of Corinth Basin is dominated by major normal faults trending east–west or east-southeast–west-northwest. Cumulative displacements over the past million years have caused more than 3 km offsets of pre-Neogene basement across these structures. Faults in the eastern Gulf of Corinth, active in 1981, were found to dip at an average 45° to focal depths of 8–11 km. Seismic lines across the eastern gulf (Alkyonides Gulf) reveal the dominance of northward-dipping normal faults along the southern margin of the basin. An asymmetric

11

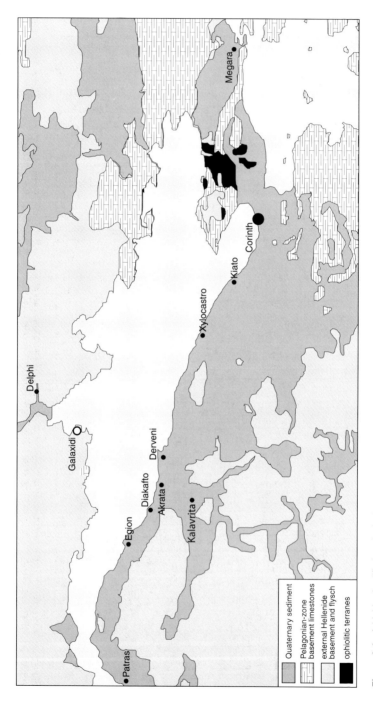

Figure 2.4 A highly simplified geological map to show chief towns, main basement zones, ophiolitic and Quaternary outcrops. Note the widespread Quaternary outcrops south of the gulf, located mostly in inactive fault-bounded basins. Greatly simplified from the 1:250 000 geological map sheet of Greece (IGME).

Quaternary sediment

Pelagonian-zone
basement limestones

external Hellenide
basement and flysch

ophiolitic terranes

Megara

Corinth

Kiato

Xylocastro

Delphi

Galaxidi

Egion

Diakafto

Derveni

Akrata

Kalavrita

Patras

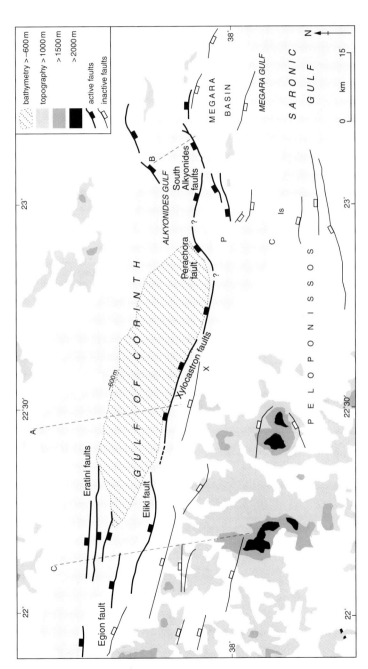

Figure 2.5 A topographic and bathymetric map of the gulf, with major onshore and offshore active and inactive Quaternary normal faults. C – Corinth city, X – Xylocastron, Is – Isthmus of Corinth, P – Perachora Peninsula. Lines A, B, C refer to the location of upper crustal cross sections in Figure 2.6.

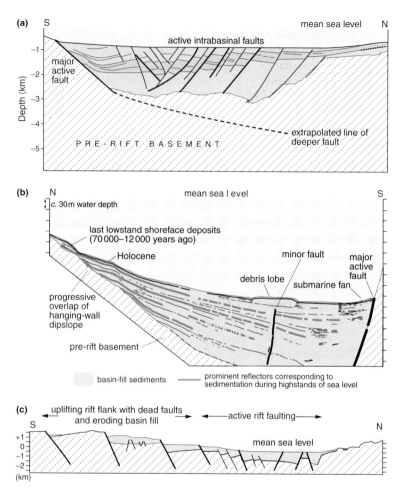

Figure 2.6 Cross sections to show the varied cross-sectional form of the active Corinth rift. **(a)** Symmetrical graben in west-central areas (after Sachpazi et al. 2003). **(b)** Asymmetric half-graben in the eastern gulf (unpublished data of the authors). **(c)** Generalized summary section in the western gulf to show both active coastal and offshore faults and inactive onshore faults bounding Quaternary sedimentary rift infills (after Moretti et al. 2003, McNeill et al. 2005a).

graben (half-graben) morphology is therefore defined (Fig. 2.6b), with the thickest Quaternary sedimentary sequences (>1 km) preserved in the southern part of the gulf. It is important to note that large south-dipping normal faults do occur along the onshore and offshore parts of the northern gulf margin in the east. One of these, the Kaparelli fault, broke surface with an earthquake of magnitude greater than 6 in 1981. However, such antithetic faults do not disturb the general asymmetry of the active half-graben. This contrasts with the situation in the western and central gulf,

14

which shows a characteristic full-graben symmetry (Fig. 2.6a), being bounded to the south by major north-dipping normal faults and to the north and within the basin by major south-dipping faults, both of which influence basin stratal dips. Intrabasinal horsts bounded on both north and south sides by active faults occur off shore in west (e.g. Eratini horst) and east (e.g. Alkyonides Islands horst) areas.

Basin structure has also been strongly affected by regional uplift across the northern Peloponnisos. The rate and cause of this uplift has yet to be fully constrained. The active southern basin-margin structures have successively stepped northwards through the Quaternary development of the basin, so that previously active basin-margin structures and depocentres to the south of active faults have become inactive and uplifted (Fig. 2.6c). The net effect has been to narrow the gulf over time. Footwall uplift and backtilting caused by fault migration are best shown by the Megara Basin, uplifted some 400 m and tilted southeast in the footwalls of the active Psatha and Skinos faults in the past million years or so. In other areas the magnitude and origin of uplift will remain indeterminate until the early sedimentary infill of the basin is adequately dated and initial relief determined. Data from raised marine-shoreline deposits indicate maximum rates of uplift of 1–1.5 mm a year over the past 200 000 years.

The Quaternary fill of the gulf is up to 3 km thick, with faults locally determining basin morphology and sediment thickness. Seismicity associated with extensional faulting has been postulated to trigger substantial slumps on submarine slopes. However, seismic reflection data (authors' unpublished data) show clearly that the only example reported after a major earthquake, in the easternmost Alkyonides Gulf (Perissoratis et al. 1984), is clearly a much older feature, perhaps dating back to the most recent glacial lowstand, for both the debris-flow lobe mapped down slope and its upslope failure scars are clearly draped by continuous sedimentary reflectors of Holocene age. Micropalaeontological evidence from offshore sediment cores (Collier et al. 2000) proves that the gulf was a lake (Lake Corinth) during the most recent glacial epoch. It is also possible that lacustrine conditions existed during previous marine lowstands if the shallow (–60 m OD) sill at the narrow entrance to the gulf is a long-lived feature.

Research history and current issues

The advent and development of plate-tectonic theory focused international geophysical attention on the Gulf of Corinth, which had long been known for the intensity and frequency of its earthquakes. In the 1970s, plate-tectonics pioneers Dan McKenzie, Xavier Le Pichon and co-workers laid

15

the firm foundations for all subsequent studies. A key was McKenzie´s (1972) recognition of the context of the Aegean volcanic arc within the overall plate-boundary framework of the eastern Mediterranean. McKenzie originally proposed 300 km displacement of Aegea, so as to minimize work involved in the Arabia/Asia Minor collision, forcing Aegea to the southwest, where Mediterranean oceanic lithosphere remained to subduct. In a later benchmark paper (1978), McKenzie identified mainland Greece, and the Gulf of Corinth in particular, as areas undergoing active continental extension. In global terms it was a superb area to study this newly recognized and distinctive phenomenon, and it was comparable to, and in many ways superior to, that of the Basin and Range in western North America; this was because of the rapidity of the extension and its maritime location, enabling palaeo-shorelines to be traced as evidence for uplift and subsidence.

Le Pichon & Angelier (1979) integrated much structural, geochronological and geophysical data to give a coherent model and timescale for evolution of an extending Aegean-arc lithosphere driven ultimately by the excess potential energy available to the Anatolian Plateau to the east. The small and highly curved Hellenic arc (radius 400 km) was related to a small circle with its pole at 38.5°N, 25.5°E. Poles of rotation for the Hellenic trench and Cephalonia strike-slip faults were positioned at 38°N, 20.5°E, consistent with mean northwest–southeast slip vectors for earthquakes averaged over sectors. As to the question of whether the length of the subducting slab was compatible with the present pattern of underthrusting, mean slab dips of about 30° gave about 170 km maximum extent of subduction. Absence of seismicity deeper than 100 km below mainland Greece and the dating of onset of volcanism in the Hellenic arc at 2.7 million years ago implied a mean rate of clockwise rotation and subduction of about 20 mm a year in western areas and 45 mm a year in the east. The crux of the paper was that rotation is taking place between the Aegean trench and Europe, with the African oceanic lithosphere retracting southwards at a rate faster than the overall extrusion rate of Anatolia, hence causing slab and trench retreat (now known as rollback, see Figs 2.7, 2.8). The overall result has been southward trench migration and overall stretching of the Aegean lithosphere. Compression north of 38°N was attributed to continental shortening, the differential slip taking place along the Cepahalonia strike-slip fault (Fig. 2.3).

It was after investigations of the intense and damaging earthquakes of 1981, led by McKenzie´s Cambridge group and co-workers from Paris and Athens, that James Jackson and Geoff King worked on the elucidation of the active and Quaternary tectonics of the eastern gulf and the Corinthian Isthmus from an integrated study of seismology, fault kinematics and

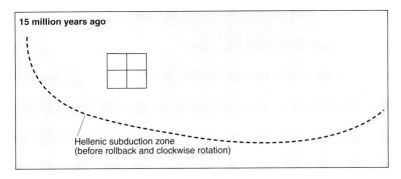

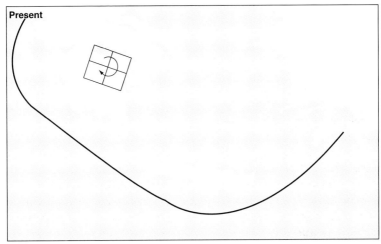

Figure 2.7 A classic illustration of Aegean tectonics. Le Pichon & Angelier (1979) used deformation grids to show Pliocene to Recent extension and clockwise rotation from southward-
translating rollback of the surface trace of the subducting African slab. We now know that extension of the overriding plate caused Mio-Pliocene detachment shear over core complexes in the
Cyclades (see Fig. 2.2) and later Quaternary to Recent high-angle normal faulting and clockwise
block rotations (see schematic rigid-body rotation of grid traces centred on the eastern Gulf of
Corinth).

surface geomorphological indicators of subsidence and uplift (Jackson et
al. 1982a, 1982b, King et al. 1985). Specifically, they developed the concepts
of:

- upper crustal continental extension occurring as fault-bounded tilt
  blocks rotate about horizontal axes (Jackson & McKenzie 1983; Fig. 2.9)
- partition of fault throw into components of footwall uplift and hanging-
  wall subsidence (Fig. 2.9)
- Quaternary migration of active faulting to cause wholesale uplift of
  formerly active tilt-block basins.

Active coastal-fault segments have subsequently been mapped along

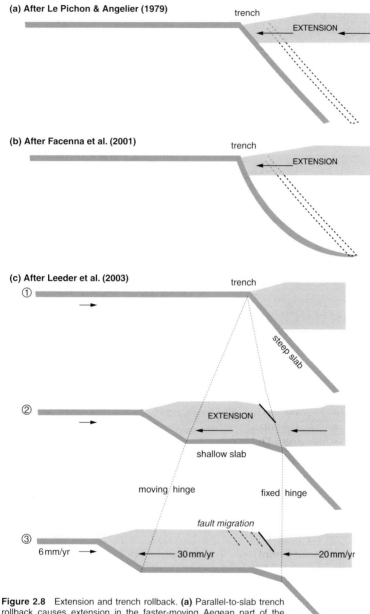

(a) After Le Pichon & Angelier (1979)

trench

EXTENSION

(b) After Facenna et al. (2001)

trench

EXTENSION

(c) After Leeder et al. (2003)

① trench

steep slab

② EXTENSION

shallow slab

moving hinge   fixed hinge

fault migration

③ 6 mm/yr   30 mm/yr   20 mm/yr

Figure 2.8 Extension and trench rollback. **(a)** Parallel-to-slab trench rollback causes extension in the faster-moving Aegean part of the advancing Anatolian plate. **(b)** In this rollback model, based on the Tyrrenhian Sea, the oceanic slab collapses oceanwards. **(c)** In this pushback model, advancing Aegea slides more rapidly over the slowly moving African slab than the rest of the Anatolian plate, deforming it along the Gulf of Corinth, with the production of flat slab under southern Greece.

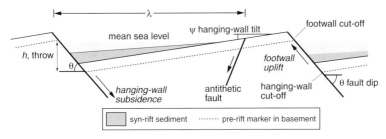

Figure 2.9 Tilt blocks and their kinematics: definition diagrams.

the entire southern rift margin, with spectacular uplifted Gilbert fan-delta deposits adjacent to inactive faults in the western gulf confirming northward migration of active faulting over time (Ori 1989). Subsequent use of seismic-moment tensor data (Taymaz at al. 1991) established first estimates of overall rates of extension across the Aegean area. These indicated very high stretching velocities (c. 60 mm yr^{-1}) and also strain rates (c. 4.10^{-15} s^{-1}), requiring correspondingly high subduction velocities (c. 100 mm yr^{-1}) in the Hellenic Trench. More recent GPS data reveal that these rates are overestimates (Fig. 2.3).

Early palaeomagnetic studies (Laj et al. 1982) found evidence for rapid vertical-axis rotations in the Aegean area after 5 million years ago (Plio-Pleistocene). These results led to development of 3-D crustal-slat and broken-slat models (McKenzie & Jackson 1986) to describe the crustal kinematics of extensional tectonics; the geometry of the deformation resembles the behaviour of crustal tilt blocks with vorticity. A simple quantitative model of such broken crustal slats reproduced most of the features of the instantaneous velocity field in the Aegean region (Jackson & McKenzie 1988). These include slip vectors and nature of faulting in eastern and western areas, sense and approximate rate of rotation of the overall extensional velocity field across the Aegean, and spatial distribution of strain rates, as evidenced by seismicity, topography, bathymetry and geodesy. More recently, Duermeijer et al. (1999) carried out palaeomagnetic measurements on very well dated Eocene to Pleistocene sediments on the Ionian island of Zakynthos (northwest Greece) that have relevance to the Corinth rift. Magnetostratigraphical and biostratigraphical constraints provided a more reliable timeframe than in previous studies. The results showed that no significant rotation had occurred between 8.1 and 0.8 million years ago, but a very rapid 21.6°±7.4° clockwise rotation occurred between 800000 years ago and the Holocene. Duermeijer et al. (2000) subsequently presented new palaeomagnetic data from the entire Aegean outer arc. The results indicated rapid clockwise late-Pleistocene rotation in the western Aegean arc (Zakynthos, as already known, and southern and western Peloponnisos) with the eastern Aegean arc (Kassos, Karpathos and Rhodes) experiencing anticlockwise Pleistocene rotations, both in good agreement with present-day rotation patterns as computed from geodetic data.

Mapping of the prominent flight of marine terraces from New Corinth west to Xylocastron (Keraudron & Sorel 1987) established that uplift along the southern rift flank was on a larger spatial scale than that appropriate to individual crustal blocks. Correlation of terrace levels with Quaternary sea-level highstands and later direct dating of corals by U–Th-series disequilibrium techniques in the isthmus and westwards (Collier 1990, Collier & Thompson 1991, Collier et al. 1992, Houghton et al. 2003, McNeill & Collier 2004, DeMartini et al. 2005, Leeder et al. 2005) proved marked spatial gradients in uplift rates, increasing from about 0.3 mm to up to 1.5 mm a year westwards. One hypothesis is that this trend reflects variations in footwall uplift caused by increasing proximity to the active offshore Xylocastro fault (Armijo et al. 1996). Moretti et al. (2003) and Leeder et al. (2003) pointed out difficulties with this idea, namely the very large fault displacements required (> 9 km) compared to those observed (4 km) in offshore seismic data, the wide extent of the uplift, and the lack of terrace tilting southwards from the fault.

Deep seismic tomographic studies (Rubie & van der Hilst 2001) reveal the extent of subducting slab under the Aegean (Fig. 2.10), the slab eventually passing through the upper mantle discontinuity at a depth of 550 km. Shallow tomographic studies (Tiberi et al. 2000) confirm earlier identification (Spakman et al. 1988) of flat slab subduction under the Peloponnisos. This has given rise to the idea that regional uplift occurs above buoyant flat slab, the uplift locally overcome by rapid subsidence in the hanging walls of major active normal faults and reinforced by uplift in their footwalls (Leeder et al. 2003). At about the longitude of the eastern Gulf, the slab dips steeply down to the locus of melting that underlies the Aegean arc.

In recent years, perhaps the most spectacular direct evidence for active crustal-extension vectors and of finite strains across the Gulf of Corinth have come from satellite GPS studies. In a landmark contribution, Davies et al. (1997) made use of a precise first-order triangulation of Greece carried out in the 1890s. Reoccupation using GPS receivers of 46 of the 93 original markers yielded estimates of the deformation of the region over the intervening interval. The observed distribution of displacements and those observed by other workers (Fig. 2.11; Briole et al. 2000, McCluskey et al. 2000, Avallone et al. 2004) can be explained by the relative rotation of two blocks, central Greece and Peloponnisos, with the rift as a broad accommodation zone between them taking up the majority of strain. Clark

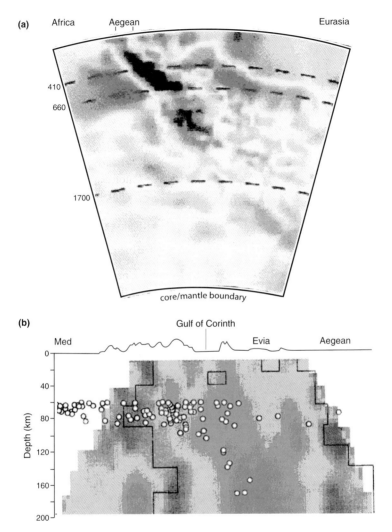

Figure 2.10 Seismic tomography sections. **(a)** The deep slab in this image extends to 1500 km depth (from Rubie & van der Hilst 2001) and has maximum greyscale variation in seismic wave velocity of ±1%, with darker shades being positive (faster). **(b)** Crust and shallow mantle image (from Tiberi et al. 2000) to show deep seismicity (white circles) and the extent of the flat slab south of the Gulf of Corinth. Maximum greyscale variation is ±5%.

et al. (1997) considered in more detail the seismic strain deficit in the western Gulf of Corinth, where the 1995 Aigion earthquake focused attention on the seismic hazard of that area. Although there have been few large earthquakes in the region during the past century, the historical and palaeoseismological record suggests that there have been many large

21

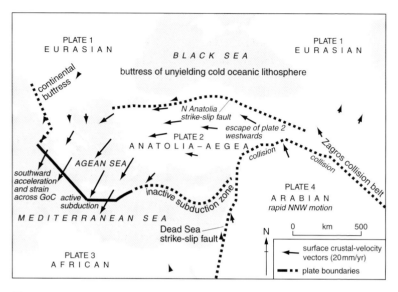

Figure 2.11 Horizontal surface velocities of the lithospheric plates making up the eastern Mediterranean and Asia Minor. Data derived from satellite geodesy platforms (GPS) averaged over a few years and stated with reference to a stationary Eurasian plate reference frame (data of McCluskey et al. 2000). Note: (1) Contrasts in velocity vectors between different plates and sharp discontinuities present across plate boundaries. (2) Evidence for systematic east to west spatial acceleration (implying crustal strain) and spin (vorticity) of the Anatolia–Aegea plate.

earthquakes there in the more distant past (see Ambraseys & Jackson 1997). The geodetic data suggest that less than half of the elastic strain in the central and western Gulf of Corinth has been released by earthquakes during this past century. In contrast, the seismic and geodetic strains in the eastern gulf are in agreement with each other. If the discrepancy between seismic and geodetic strains in the west that has accumulated during this century should be removed in earthquakes, the moment release would be equivalent to several earthquakes of magnitude greater than 6.5.

Interestingly, the north-northeast–south-southwest extension that has caused the major geological features noted above has not systematically thinned the crust of the Corinth rift, as would be expected from simple or pure shear models of crustal extension. Recent gravity and seismic data (Zelt et al. 2005) suggest that crustal thickness decreases meridionally from the 40 km or so in the west to less than 30 km in the eastern gulf, directly under the Perachora Peninsula. The detailed pattern of crustal thickness suggests that there was much pre-rift crustal-thickness variation and that the mismatch in places between current thickness and topography (especially around the Perachora Peninsula and Gerania Mountains in the eastern rift) suggests a lack of local isostatic compensation.

Current research issues

The recent literature points to the following research issues:
- timing, origin and rates for regional uplift of the southern rift flank
- timing and mechanisms for progressive northern migration of fault death along the southern rift flank
- delimitation of offshore normal faulting and its history in relation to onshore faulting
- existence and role of low-angle detachment faulting during extensional history
- the kinematic imprint of major antithetic faults along the northern rift margin
- partition of co-seismic fault displacement between footwall uplift and hanging-wall subsidence and its evolution in post-seismic intervals
- timing of vertical-axis crustal-block rotations with respect to rift evolution
- radiogenic and palaeomagnetic dating of Quaternary stratigraphical and sedimentological events in order constrain tectonic evolution
- evidence for and chronology of Quaternary climate change and its influence on hydrology, geomorphology and sedimentology
- longevity and hydrodynamics of Lake Corinth in glacial epoch lowstands
- rates of river incision and the magnitude of sediment fluxes in relation to changing palaeoclimate, tectonics and sea level.

Theoretical field geologists and geophysicists should note that none of the above research agenda will be achieved without adequately detailed geological and offshore geophysical fieldwork.

Further reading

Very useful compendiums of papers on Greek geology and its wider context within the east Mediterranean are edited by Dixon & Robertson (1984) and Robertson & Mountrakis (2006). The evolution of the Aegean within the context of the story of NeoTethys is told by Robertson et al. (1991). The discovery paper on Cyclades core complexes is by Lister et al. (1984) and the most recent syntheses may be found in Ring & Layer (2003) and Kumerics et al. (2005). The puzzling present crustal structure of the rift is outlined by Zelt et al. (2005). A broad overview of the Corinth rift is given by Armijo et al. (1996), although the tectonic model adopted for rift margin faulting is in error. Moretti et al. (2003) point out the importance of offshore structures in understanding rift tectonics, an approach emphasized by the results of McNeill et al. (2005). Other offshore studies of submarine geology, geomorphology and geological structure include those by Brooks & Ferentinos (1984), Ferentinos et al. (1988), Papatheodorou & Ferentinos (1993), Sakellariou et al. (1998), Leeder et al. (2002), Stefatos et al. (2002) and Sachpazi et al. (2003).

Chapter 3

The south Alkyonides Gulf faults

MIKE LEEDER

As noted previously, the normal faulting and crustal deformation in the eastern Gulf of Corinth were first brought to light by field and seismological investigations after the 1981 earthquakes. Since then the area has received widespread attention from many geological and geophysical points of view (see Further reading). As a result of all this attention, the normal faults in the eastern gulf (Fig. 3.1) are undoubtedly some of the most closely studied examples in the world; lessons learned from their behaviour have illuminated many palaeotectonic and basin-analysis studies.

Key objectives

In this itinerary we address the following issues:
- scale, morphology and surface displacements associated with active, basin-bounding fault systems
- depositional and erosional features along a faulted bajada
- palaeoseismology and landscape growth.

Initial odometer readings are taken from zero at the Loutraki town fountains on the north end of the main street (37°56'45.6"N, 22°58'40.8"E; elevation 9 m). Loutraki is a spa resort nestling at the base of the Gerania Range (pronounced "Yerania", meaning "stork"), mainly Mesozoic platform limestones here but with large outcrops of serpentinites farther east and west. It is built partly on raised coastal deposits of most recent interglacial age and on the distal portion of the Smarpsi alluvial fan, whose drainage catchment includes much of the steep southwest slope to the range. The fan provides characteristic serpentinite, limestone and chert clasts to nourish the modern beach system of the northwest shoreline to the isthmus. Evidence gained at stop 6.2 suggests that this situation has prevailed for at least 300 000 years.

Follow the main street north, leading eventually (10 km) to the village

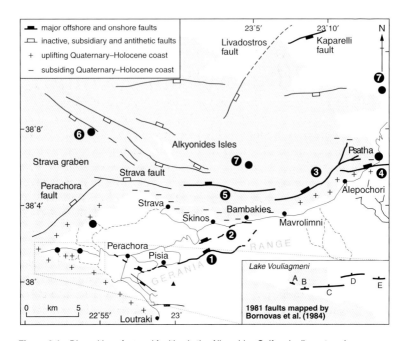

Figure 3.1 Disposition of normal faulting in the Alkyonides Gulf and adjacent onshore areas, eastern Gulf of Corinth and location map for the itinerary in this chapter. Onshore 1981 fault breaks after Jackson et al. (1982a), with eastern extent of Skinos fault after Collier et al. (1998). Offshore faults mapped by Papatheodorou & Ferentinos (1993), Leeder et al. (2002) and Stefatos et al. (2002), from whom the fault names are taken. Numbered faults are 1. Pisia fault, 2. Skinos fault, 3. East Alkyonides fault, 4. Psatha fault, 5. West Alkyonides fault. Filled circles 6–8 are the favoured locations for the epicentres to the three 1981 earthquakes (after Taymaz et al. 1991). The status of uplifting and subsiding coastline data is from Jackson et al. (1982a), Hubert et al. (1996) and Leeder et al. (2002, 2003). Inset shows faults A–E in the western Perachora Peninsula mapped by Bornovas et al. (1984).

of Perachora. After climbing via broad sweeping bends with roadcuts exposing Mesozoic limestone and thrust contacts with serpentinite and chert (e.g. 4.3 km) note a steep gorge to the left (5.3 km) whose drainage feeds the fan delta on the southern side of Agriliou Bay (see stop 4.6). Then, after a series of very sharp hairpins by the Panorama taverna and before the straight stretch leading to Perachora village, take the right fork at 6.2 km to Agios Geracimoy monastery, with the arête of the Gerania Range prominently in view. After 600 m, turn right into the large monastery drive with the prominent white cross. There is ample coach-turning space in the monastery car-park. There are no particular safety issues in this locality. (Coordinates: 38°00′26.4″N, 22°56′35.5″E)

Itinerary

Stop 3.1 Scene setting and viewpoint The vantage point, at about 315 m elevation, below the western tip to the Gerania Range offers spectacular views to the south and west and is a good place to set the context of eastern gulf itineraries (Fig. 3.1). The view, best taken early (a haze often develops in summer), stretches west across the Bay of Corinth and to the southern rift flank. Due west is the Perachora Peninsula and Lake Vouliagmeni (Ch. 4). Almost the whole of the view has been uplifted above mean sea level during the late Quaternary. In the immediate foreground to the north is the village of Perachora, which, together with Pisia village to the east, bore the brunt of the 1981 earthquake; both were mostly destroyed, with loss of life. Note the deforested slopes around Perachora village, the effects of devastating fires that swept the peninsula in the late 1990s.

Stop 3.2 Pisia village for 1981 and precursor fault scarp Drive out of monastery car-park back down to the main Perachora road (zero odometer). The straight stretch crosses a prominent flat interior drainage basin at about 250 m elevation fed by the axial drainage that receives the combined northward flow of several tributaries from the western Gerania Range. At the beginning of the straight (1.0 km), within a large olive grove, scarplets (throws of a few centimetres) almost at the western extremity of their distribution formed during the 1981 earthquakes were mapped shortly afterwards (Jackson et al. 1982a). They cut the alluvium of a prominent alluvial cone marking the outfall of a major axial (i.e. fault-parallel) drainage. Further ground cracking was later recorded several kilometres to the west (Bornovas et al. 1984), although the significance of these westernmost fractures in the larger scheme of things is controversial (cf. Morewood & Roberts 1999, Leeder et al. 2005). Bear right (2.3 km) at the fork, with the road descending left to the Heraion archaeological site (leading to the itinerary in Ch. 4) and take the tortuous street through up through the village, following the occasional signs to Pisia and Skinos, and eventually turning east with the spectacular vista in front of the Gerania Range (maximum elevation 1150 m). Then drive along the foot-hills to the hamlet of Pisia, which is visible at the base of the steepest range front slope ahead at 5.3 km.

In Pisia (15.4 km), park in the recently widened road some 200 m west of (i.e. before reaching) the church. Walk to the church via the narrow (passable by coach) main street. A memorial tablet in the wall commemorates the casualties of 1981. Walk 400 m south from the church along a recently landscaped by-road that is fly-posted frequently to the Platanos taverna. Take in views of the impressively steep range-front scarp as you

27

do so. You might ponder on the number of normal faulting events it might take to raise such an edifice some 1000 m above sea level. Walk across the exterior patio to steps at the left side adjacent to the kitchen (the owners are accustomed to this) leading to the base of an abrupt scarp, which is the object of our attention (Fig. 3.2). Safety helmets are required. Take care to avoid the loose talus slopes on either side of the main exposure. In the event of rain, do not climb the smooth limestone scarp, because the descent becomes embarrassingly slow. (Coordinates: 38°00′57.7″N, 22°59′20.6″E)

The large, corrugated (mullioned) normal fault-scarp surface is in Mesozoic limestone. It has been exposed by erosion at some time past (but pre-1981) of talus gravels seen on either side of the exposure. The 1981 ground displacement is clearly shown by the abrupt top to a paler, less weathered and less lichen-encrusted zone at the base of scarp. The fault strike, dip and throw (vertical displacement) are 90°/50°/1.0 m respectively. It is impossible to determine here the partition of displacement in 1981 between footwall uplift and hanging-wall subsidence. The mullioned features marking the long-term slip azimuth over the larger scarp (Fig. 3.2) run slightly oblique to the true dip of the scarp (the rake is the clockwise angular deviation from horizontal, about 75–80° here); the fault scarp has thus experienced repeated dip, with a small western oblique component. Looking up at the whole scarp and extrapolating over the range front where the scarp has undergone erosion, we can infer episodic growth by faulting of the Gerania Range front (maximum elevation c. 1.3 km) above the gulf (maximum depth 400 m with c. 1 km of sedimentary infill). If the 1981 displacements are typical (see stop 3.5, Skinos fault), about 2500–3000 earthquakes at c. 1 m increment per earthquake must have been responsible. Knowing the average recurrence interval of faulting (about 350 years for the Skinos fault to the east), we may estimate the approximate time taken for development of range-front elevation as about a million years. This admittedly crude estimate is nevertheless astonishingly close to other independent estimates (see Ch. 5).

Three rupture events occurred during the 1981 Corinth earthquake sequence. The first two occurred during the night of 24/25 February and ruptured north-dipping faults, the first along the rupture we see part of here. The third event occurred on 4 March and ruptured the south-dipping Kaparelli fault on the north side of the gulf, defining what is termed an antithetic fault scarp (i.e. dipping opposite to the polarity of the major southern basin-bounding structures). Revised magnitudes for the individual earthquakes are 6.7, 6.4 and 6.4. Two major parallel faults, the Pisia and Skinos faults, were recognized from mapping of the surface breaks on the south side of the gulf. Each fault comprised surface rupture strands 0.25–7 km long, separated by short gaps (up to 1 km). Surface breaks with a

Figure 3.2 Pisia fault scarp with prominent mullions and the line marking the boundary between less- and more-weathered limestone, ground level before the 1981 displacement, at about head height. People are Greg Mack, Lela Mack and Marta Pérez-Arlucea.

maximum 1.5 m throw and slip vectors generally to the north-northeast were mostly coincident with basal slopes to major linear escarpments. The Pisia fault linked to the first earthquake had mean displacements of 0.9 m. The second earthquake associated with the Skinos fault (stops 3.3, 3.5) caused a mean displacement of 0.3 m.

Stop 3.3 Skinos headland foreshore viewpoint   Rejoin coach, reset odometer to zero at the church and drive east to Skinos along the winding road at the foot of the Gerania Range bajada, parallel to the Pisia fault. Soon (2.6 km)

29

Figure 3.3 Panorama of the Skinos bajada, with its prominent alluvial fans and the great double escarpment of the Gerania Range. The Skinos fault bounds the nearest range front; the eastern tip of the Pisia fault bounds the higher of the two scarps. The steep decline of the latter in the centre of the photo is not attributable to the Pisia fault tipping out, but to a change in basement lithology from limestone to easily weathered serpentinite.

we begin to descend, with the first open views of the Alkyonides Gulf at 3.0 km. Exposures of an older fault scarp occur between 3.9 km and 4.4 km. At a sharp hairpin bend (6.6 km) there are fine views of an excavated exposure in the Pisia fault in an old quarry across the valley. If required, this can be accessed by a track that left our route to the right at 5.9 km. Continue to descend to another steep east–west escarpment marking the Skinos fault. From 6.8 km to 8 km the roadside cuttings are in Mesozoic limestone basement, which is intensely fractured, with Neptunean dykes, carbonate-filled veins and cave flowstone, which are all common locally. The irregular karstified surface limestone is overlain by a prominent reddish-brown terra rossa soil. Red cherts and serpentinites become more common as we descend the lower slopes, the latter being nicely exposed at 9.5 km. There are fine views east now along the coastal escarpment and of the Alkyonides Gulf to the north. At 10 km, ahead, the overlapping Pisia and Skinos faults define a stunning double escarpment on the high skyline. At 10.5 km, note the spectacular gorge cut into the limestone footwall to the Skinos fault. At the base of slope we drive on the prominent alluvial fan that has formed at the confluence of several such incised drainages, and at 11.6 km take the left fork towards Skinos village centre, where we pass (13 km) the taverna whose patio subsided a few decimetres during the 1981 earthquake. Fork right, following signs for Alepochori, past villas, shops, a petrol station and church, to a left junction at 14.1 km with a

metalled minor road up a small hill. Follow around to the right at 14.5 km (accessible by coach and where the coach can turn) to the beach on the southeast margin of the Skinos headland and view the illuminating scene along the south shoreline of the Alkyonides Gulf.

There is parking along the beach access road. There are no particular safety issues to worry about in this locality. (Coordinates: 38°03'26.6"N, 23°02'31.2"E)

We stand below the imposing Gerania Range front (Fig. 3.3), in the subsiding hanging wall to the Skinos fault. The vista presents a perfect view of a footwall range drained by steep-gradient streams. In front is the coastal bajada, comprising coalesced alluvial fans, talus cones, fan deltas (alluvial fans that debouch directly into the sea), coastal lagoons, marshlands and beach-barrier spit and beach shorelines. The alluvial fans have their apices located up slope from the line of the Skinos fault scarp, the surface of each one being incised to a greater or lesser degree. The toes of these fan deltas are mostly cliffed, with vertical faces up to 17 m high. Because of the presence of the cliffs, fan channels are deeply incised at their seaward exits, the degree of incision decreasing up fan. Sediment is supplied to the coastal tract by the streams that drain the basement hinterlands of the fault footwall and by reworking of the cliffed toes of the proximal fans by wave action. The basement lithology is predominantly Mesozoic limestone in this western area and serpentinite in the eastern catchments (see stop 3.4). From our previous discussion (see also Ch. 5) it is likely that the bajada system has been present here for about the past million years or so. Offshore seismic reflection surveys reveal a 2–5 km-wide ramp located on the footwall of a prominent offshore fault, the West Alkyonides fault, thought to be currently dormant. Seismic profiles also reveal that both sub-aerial and sub-aqueous bajada deposits and the offshore fault foot-wall are subsiding in the hanging wall to the active Skinos fault along this stretch of the coast. This scenario changes rapidly to the east (stop 3.6).

Stop 3.4 Section through alluvial fan and view of talus cone   Rejoin the main road to Alepochori, zero the odometer and drive east. Ahead of you are prominent cliffed fan deltas at the base of the footwall scarp. We stop at the roadside quarry at 1.4 km. For parking, pull in the quarry entrance. Beware of the traffic coming around the sharp bends. (Coordinates: 38°03′10.3″N, 23°03′19.4″E)

In the quarry we see a section (Fig. 3.4) that illustrates the internal archi-tecture and lithologies present in one of the constituent alluvial fans and fan deltas that make up the Skinos coastal bajada. The majority of the section comprises crudely stratified, lenticular-bedded, grain-supported, open-framework sub-angular to sub-rounded serpentinite gravels. These are derived from a catchment draining the extensive ophiolite outcrops that make up the higher Gerania Range at this locality. Note that the imme-diate fault scarp is in Mesozoic limestones. The sediment has the charac-teristics of streamflow deposition rather than those of more viscous debris flows, where matrix support by fines might be expected. The periodic streamflow deposits define decimetre- to metre-thick lenticular wedges, probably lobate in 3-D form. They are accentuated by intercalated buff-grey fine silt, probably ultimately of windblown origins, but reworked from higher in the fan as the silt has intercalated sand and granule string-ers and it often occurs in shallow lenticular channel-like forms cut out laterally by the coarser sediment. A prominent, laterally continuous, reddish-brown terra rossa palaeosol developed in one of these is about

Figure 3.4 (below, opposite, and detail above)   Quarry section through alluvial fan deposits. Mike Leeder, Martin Brasier and Alex Brasier are viewing the palaeosol, which is at shoulder height.

0.3 m thick, brick red in its upper 10 cm, and must record a longish break in deposition over much of the fan. A ^{14}C date from disseminated organic matter from the soil gave an early Holocene age, 7620±40 years ago. Because of the nature of the disseminated carbon in the sample, the date must be regarded as a mean age, and the onset of soil formation may have been somewhat earlier, the exact date depending on local sedimentation

rates. It is likely that the soil profile may record stabilization of catchment landscape by Mediterranean forest, which achieved climax conditions as Holocene sea level was approaching its maximum highstand at about this time in the gulf (see Collier et al. 2000). The underlying gravels are therefore likely to be youngest Pleistocene, probably dating from during or after the most recent glacial maximum.

The 4 m of overlying Holocene gravels are also deposited as lobes and in channel forms, a particularly prominent example cutting out the palaeosol in the central part of the exposure. They indicate renewed deposition that must mark a fundamental change in catchment hydrological and runoff characteristics after the stabilization recorded by the palaeosol, perhaps aided by mid-Holocene climate change or Bronze to Iron Age deforestation, or both. The computed conical volume of the gravels younger than 7620 years old extrapolated out from the fan apex to modern sea level gives a total sediment volume of some $0.006 \, km^3$ and a mean mechanical denudation rate from the catchment since that time of some 0.26 mm a year. The fan has more recently been progressively truncated and cliffed by marine erosion and incised by its active channel as the hanging-wall surface subsided during periodic earthquakes along the Skinos fault (see stop 3.5).

Stand back from the quarry face and look up to the scarp, with its striking talus cone to the south (Fig. 3.5). The Mesozoic limestone forming the footwall here is intensely karstified. Such precipitous cave-ridden outcrops are inherently unstable during earthquakes, and their collapse, by toppling failure, causes the descent and comminution of talus along chute-like conduits. These form striking downslope-coarsening talus cones, which have debouched with little runout onto the surface of the alluvial fan. Rare limestone boulders within the serpentinite gravels in the quarry section undoubtedly owe their origin to frontal bouncing after similar gravity-flow processes in the past. Evidently, the periodic collapse of the karstified limestone footwall has led to both maintenance of the steep scarp and its concomitant parallel retreat away from the active fault line. At least two of the talus cones along this stretch of footwall formed by slope failure after the 1981 earthquake event, according to the evidence from aerial photographs (comparing 1944 and 1987 vintages) and conversations with local residents.

Before leaving, note the adjacent coastal marshlands and lagoons that define interfan areas along the subsiding coastal bajada. As well as providing valuable wetland ecological niches, they define tempting targets for sediment coring to aid recognition of previous coastal inundations (including possible tsunami events) following episodes of co-seismic hanging-wall subsidence.

Figure 3.5 Talus cones abutting the alluvial fan and limestone karst footwall.

Stop 3.5 Bambakies fan, 1981 fault breaks and palaeoseismological trenches
Continue east on the road to Alepochori. At 2.3 km there are serpentinite
outcrops, with limestone briefly again from 2.6 km to 3.3 km. The Skinos
fault bounds these outcrops, but is visible as a fault plane only in the more
resistant limestones, such as at 3.2 km (38°03'17.5"N, 23°04'32.1"E) where
the fault strikes 244° with a 55° dip north. Reach Bambakies alluvial fan at
about 3.7 km, the hamlet ahead built somewhat optimistically in view of
flooding potential on the surface of the alluvial fan. Park in the layby near
the church. There are no particular safety issues in the locality. (Coordinates: 38°03'23.0"N, 23°04'32.1"E, elevation 83 m)

In the 1990s a church was built, with not a little faith, *in* the active channel of the alluvial fan. The *ad hoc* flood-relief excavations to the east reveal
good sections through alluvial-fan serpentinite gravels comparable to
those seen at stop 3.4. A weak iron-rich palaeosol at about 4 m depth has
yielded a radiocarbon age on disseminated organic carbon of 2550±40
years ago, consistent with a local pause in upper Holocene deposition
(Classical Greek period). Good sections are also seen in the deeply incised
seaward cliffs, where extrapolation of the longitudinal fan surface slope to

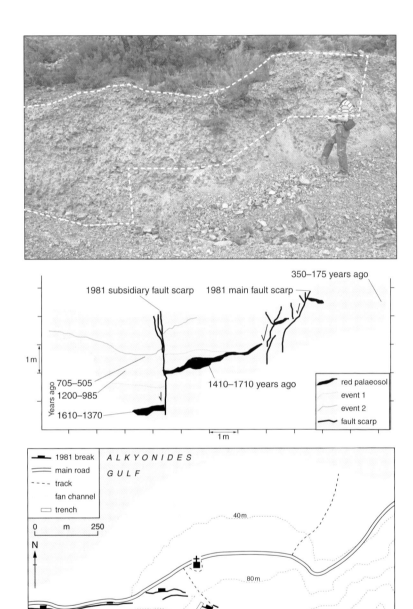

Figure 3.6 Bambakies alluvial fan: Skinos fault palaeoseismology trench section and interpreted sketch (after Collier et al. 1998). Greg Mack provides scale.

present sea level gives a figure of about 150 m for the amount of fan retreat since the beginning of Holocene cliffing by marine erosion.

Walk back 25 m down the road and then turn left up slope on the steep track over the fan surface for about 150 m, where the Skinos fault scarp becomes visible at the front of limestone outcrops to the west. Approximately in line with this scarp, walk 10 m west to encounter a trench excavated by a joint Italian–British team (Collier et al. 1998) in the mid-1990s (Fig. 3.6; location 38°03′16.7″N, 23°04′33.8″E, elevation 83 m). The trench clearly shows the subsurface fault trace, serpentinite clasts rotated into the fault plane and traces of the faint immature Fe-stained palaeosols used for AMS radiocarbon dating. The 1981 surface break occurs to the right of the trench.

In contrast to the Pisia fault locality (stop 3.2), 1981 surface faulting here occurred in Quaternary alluvial-fan sediment. Topographic profiling across the scarps at various positions over the fan surface yields a mean vertical displacement for the 1981 event of about 0.7 m (locally as much as 1.3 m), decreasing rapidly eastwards to zero at the fault tip on the eastern extremity of the fan. In the central part of the fan, the 1981 free face (0.5–0.7 m high) is located along a pre-existing scarp, which displaces the Holocene fan surface up to 5 m. This scarp is faced by an antithetic scarp up to 1.5 m high, where 0.1–0.15 m of surface faulting occurred in 1981. The main precursor scarp represents the cumulative effect of repeated Holocene surface-faulting events on the Skinos fault segment. On the basis of the mean net height of the scarp and assuming an age for the beginning of the latest phase of fan-lobe deposition of about 2550 years (i.e. younger than the mean age of the dated palaeosol by the church noted above), a minimum 2 mm-a-year late-Holocene vertical slip rate can be inferred for this section of the fault. Similarly, assuming the 1981 event to be characteristic of this fault segment and considering the 5 m-high precursor Holocene scarp, we need about seven 1981-type events of mean 0.7 m displacement to construct it. A maximum late-Holocene average recurrence interval in the range 350 years is thus required. Detailed analysis of ^{14}C dates of palaeosols in stratigraphical units dating from late Roman Empire times cut by the excavated fault gives evidence for a total of six previous fault rupture events, which are comparable to the 1981 deformation in terms of displacement, yielding mean vertical displacement rates of 0.7–2.5 mm yr^{-1} and a more precise estimate of mean recurrence interval of about 330 years. For 45°-dipping normal faults, these rapid vertical displacement rates equate to horizontal displacement. They are significantly slower than horizontal rates of continuing steady-state crustal extension determined by geodetic surveys at the longitude of the outer Alkyonides Gulf, 6±2.7 mm a year over the past hundred years. Either the geodetic rate decreases in the

20 km interval eastwards to Skinos, or the displacement is also taken up on offshore faults, or the short-term aseismic steady-state rate overestimates the longer-term co-seismic record from palaeoseismology. Either way, calculations of seismic energy released from the 1981 earthquakes compared to current rates of accumulating strain indicate that no significant seismic-stress deficit exists today in the area. We return to estimates of longer-term slip rates below.

Stop 3.6 The marine isotope stage 5e marine terrace at Alepochori Carry on east along the narrow coastal road to Alepochori in the subsiding hanging wall to the Skinos fault. Note the frequent weathered outcrops of serpentinite on the way. As we pass the hamlet of Mavrolimni ("dark lake"), with its deep lagoon and reinforced beach barrier, the Skinos fault passes off shore (5.8 km). For the next 6 km we are on an uplifting shoreline in the footwall of the Skinos fault and that of a neighbouring offshore structure, the east Alkyonides fault. At 8.5 km a sea stack is visible at a promontory and behind lies the coastal resort of Alepochori. At a prominent hairpin bend by the stack (9.5 km) we leave the Mesozoic basement rocks seen along the route so far and pass into topographically more subdued Pleistocene infill to the abandoned Megara Basin (see Ch. 5), whose soft and relatively easily eroded fill forms the lower ground to the east of the coast road. The prominent sea stack (Fig. 3.7) formed in a toppled boulder of Pleistocene breccio-conglomerate has a double-notch rim at its base, the upper one being clearly inactive and older, emphasizing the uplifting nature of the coastline (see stop 3.7 and Ch. 4 for more on coastal notches as palaeo sea-level indicators).

Farther west, the landscape becomes spectacularly rugged badlands with deep gorges (Ch. 5). The low ground bordering the coast comprises Holocene alluvium and marine terraces with inner edges at 10 m and about 40 m elevation cut into the older Pleistocene sediments.

In the western outskirts of Alepochori take the road right at Poseidon camping (13.7 km). After ascending for about 500 m, at 700 m the road levels at a left-hand bend, crosses a small valley to the left, and bounds open fields with scattered olive trees. Turn here using the field and park up. Walk back a few tens of metres inside the field boundary, with the fence section on your left, and descend into the shallow valley bottom. Exposures may be found on the right bank for some 50 m down valley. There are no particular safety issues in the locality. (Coordinates: 38°04′53.4″N, 23°11′01.1″E)

The terrace top we have just descended is at 32–34 m elevation, rising gently to about a maximum of 40 m up slope of the parking place. The scrappy exposures and their talus cones reveal well sorted marine beach

Figure 3.7 The sea stack seen to the west of Alepochori village, located in Pleistocene talus breccias close to the southwest margin of the Megara Basin (see Ch. 5). There are two notch lips, a lower modern one and an upper level, the latter indicating sudden footwall uplift during a co-seismic event, probably along the East Alkyonides fault located several hundred metres off shore.

sands with gravel layers containing well rounded pebbles. They are noteworthy because of the occurrence of abraded stems of the coral *Cladochora caespitosa* towards the top of the exposure. Two coral samples from the deposits yielded U-series dates of 90 000, 130 000 and 150 000 years (Leeder et al. 1991, Dia et al. 1997), the first two corresponding to the most recent interglacial. Given the likely 40 m elevation of the terrace inner edge, and the existence of a lower terrace at about 12 m elevation, the middle date corresponding to marine isotope stage (MIS) 5e is the more likely age (150 000 years ago was a lowstand). An MIS 5e age would imply mean uplift rates since that time of about 0.3 mm a year. The lower terrace is probably that produced during either the subsequent 5c or 5a highstand, but there is no radiogenic data with which to decide.

Stop 3.7 Southern faulted margin to Psatha Bay: coastal fault scarp with prominent notched base Carry on to the coastal resort of Alepochori, noting the spectacular badland topography to the east, cut by steep drainages eroding Megara Basin infill. The drainage divide of the Gerania Range makes a marked detour inland here in response to very high erosion rates at this end of the Megara Basin. The drainages responsible for the accelerated erosion enter the Alkyonides Gulf along this stretch of coast and are responsible for deposition of a base-of-slope submarine fan descending to the 360 m-deep basin plain that occurs off shore in the immediate hanging wall to the offshore East Alkyonides fault (Leeder et al. 2002). It is this active fault (the fault tip displaces Holocene sediments) that is responsible for the coastal uplift we have witnessed since stop 3.5. Further evidence for uplift is seen in the frequent occurrence of very large slabs of cemented gravels that form raised beachrock along this stretch of the coast (see also stop 4.5). Drive through Alepochori, passing the right turn to Megara in the middle. Drive along the coast to the prominent coastal cliffs overlooking Psatha Bay. A coach can turn at the roundabout at the end of the scarp. Park anywhere along the wide road, preferably on the seaward side in view of the topography. The entire scarp frontage is accessible. Safety helmets are required (prayers may help) and beware of traffic speeding down this straight in summer. (Coordinates: 38°05'48.9"N, 23°12'11.3"E)

Here the new road constructed in the 1990s follows the foot of a 1.3 km-long east–west trending footwall scarp in Mesozoic limestone. The scarp is attributable to the Psatha fault, a west-southwest–east-northeast structure also mapped c. 1.5 km off shore as the Holocene-active western continuation of a 7.7 km long onshore scarp. There was no recorded displacement here in 1981. Seismic coverage suggests, but line spacing is too coarse to prove, that the Psatha fault tip ends off shore before that of the East Alkyonides fault (i.e. the faults are separate). Where uninterrupted by rockfalls, the lower 20–30 m of the face is smooth, with locally strong oblique slip indicated by mullions. It was originally proposed that the fault lost displacement rapidly to the east, an observation not supported by recent detailed mapping. The upper 50 m of the scarp has receded through toppling collapse. A screen of calcite cement often covers the exposed fault plane and strongly raked mullions in many places. This is a late Pleistocene feature, it is cut by the prominent basal Holocene notch (Fig. 3.8). As we shall discuss further in Chapter 4, basal notches cut in limestone may be attributable to corrasion by wave action, biological erosion or chemical solution. This example may have a significant contribution from the last cause because of dissolution from strong cool freshwater springflow that used to be observed at the base of the cliff before construction of the coastal road in the early 1990s (Fig. 3.8): the mix of cool fresh water and warm sea water gives nonlinear undersaturation effects, enhancing dissolution. The notch itself is a prominent 2 m-high uplifted feature previously used (Leeder et al. 1991) to estimate a mean footwall

Figure 3.8 Views of the Psatha footwall scarp with its magnificent basal notch. **(a, above)** The 1980s vintage natural state before the new coastal road ruined the ambience (ace climber RLG for scale). **(b, overleaf)** The accessible but lamentable present day scene of the same notch (an older RLG and student for scale).

uplift rate for the Psatha fault of about 0.3 mm a year since the local Holocene marine highstand abutted the Psatha fault scarp at about 7000 years ago. This rate is a minimum value, as the notch would have required some considerable time to form following the local highstand; hence we may be looking at more recent Holocene displacements of considerable proportions. Seismic data taken a few hundred metres off shore indicate

41

displacement, by 8–10 m, of a prominent reflector interpreted as an indu-
rated lowstand soil horizon, over which the early Holocene (*c.* 12 000 years
old) transgressive deposits lie. The mean Holocene displacement rate on
the Psatha fault close to its tip is thus 0.7–0.8 mm a year, with a footwall
uplift:hanging-wall subsidence ratio of some 30 per cent.

Other key evidence for footwall uplift of the scarp occurs in the form of
raised beach gravels preserved in partially exhumed fissure fills that define
Neptunean dykes (Fig. 3.9). These are locally cemented by the vadose
calcite cement screen noted above. The dykes may be most clearly seen at
17.2 km, some 20 m east of the roadsign facing east and warning of a bend
ahead. The deposits comprise well sorted and rounded polymict pebbles
(serpentinite, chert, limestone) cemented by fibrous low-Mg calcite. The
latter is ^{14}C-dated to 24 690 ± 170 years ago, approximately the most recent
glacial maximum. The cement has an oxygen-isotope composition consistent with precipitation from meteoric water. The highest gravel is at 10.2 m
above mean sea level, about 2 m beneath a discontinuous ledge present
along the western margin of the fault footwall, which may represent the
inner edge and small palaeocliff, down which the beach gravels fell into
the fissures. We cannot date the gravel directly but the youngest plausible
age for its formation is MIS 5a at about 83 000 years ago when sea level was
about 10 ± 4 m lower than it is today. This gives a maximum estimate for
mean footwall uplift rate from 83 000 years ago to present as 0.24 ± 0.05 mm
a year, a value similar to the mean Holocene rate.

Figure 3.9 Neptunean dyke at Psatha.

Finally, drive to Psatha Bay, noting at the roundabout the partly quarried alluvial fan and talus cones that abut the eastern part of the fault scarp. Proceed to the northern peninsula bounding the bay for panoramic views (Fig. 3.10) of the coastal traverse between stops 3.6 and 3.7.

Further reading

The area of the itinerary is discussed from the point of view of seismology by King et al. (1985), Taymaz et al. (1991), Abercrombie et al. (1995), Hubert

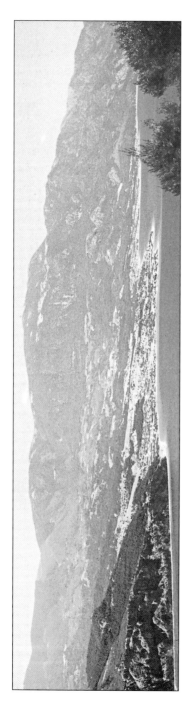

Figure 3.10 Panoramic view looking southwest to show the Psatha fault scarp (stop 3.7; foreground left), coastal terraces above the village of Alepochori (stop 3.6), northeast termination to the Megara Basin (centre, lower ground) and the 1300 m-high Gerania Range as backdrop.

et al. (1996) and Hatzfeld et al. (2000). Palaeoseismology from ancient earthquake records is by Ambraseys & Jackson (1997) and, more directly, through trenching along the Skinos fault, by Collier et al. (1998). Offshore marine geology and geophysics have thrown a great deal of light on onshore processes, and much data is presented in Perissoratis et al. (1986, 2000), Papatheodorou & Ferentinos (1993), Sakellariou et al. (1998), Leeder et al. (2002), and Stefatos et al. (2002). The tectonic and structural evolution of the area was first elucidated by fieldwork following the 1981 earthquakes in benchmark papers by Jackson et al. (1982a,b) and King et al. (1985). Subsequent work simply builds on these foundations e.g. Vita-Finzi & King (1985), Jackson & White (1989), Collier et al. (1992), Roberts & Stewart (1994), Armijo et al. (1996), Roberts (1996), Roberts & Koukouvelas (1996), Leeder et al. (2003, 2005), Morewood & Roberts (1999) and Cowie & Roberts (2001). Onshore and offshore sedimentological and geomorphological studies relevant to wider tectonic, structural and palaeoenvironmental issues are in Bentham et al. (1991), Leeder et al. (1991, 1998, 2002, 2005) and Gawthorpe et al. (1994).

Chapter 4

Perachora Peninsula to Agriliou Bay

JULIAN ANDREWS, CLIVE PORTMAN, PETER ROWE

The Perachora Peninsula lies in a pivotal geographical position, defining a large-scale left step between the active faults of the central and eastern parts of the Corinth Rift and the Alkyonides sub-basin (see Fig. 3.1). Based on field observations and interviews with local inhabitants, Vita-Finzi & King (1985) suggested that much of the peninsula underwent subsidence because of the effects of the 1981 earthquakes, a conclusion at odds with later observations in the area (Hubert et al. 1996). Regarding Holocene to Late Quaternary deformation, several authors (e.g. Vita-Finzi 1993, Pirazzoli et al. 1994, Dia et al. 1997, Leeder et al. 2005) subsequently proved persistent Holocene to late Quaternary uplift by radiogenic dating of marine-fossil assemblages in raised marine-shoreline deposits. Some of the Late Quaternary carbonate sediments on the peninsula are spectacular, particularly the large microbial bioherms that crop out on the north-west shoreline between Cape Heraion and Sterna. Moreover, there has been controversy concerning the origin, age and palaeoenvironmental significance of these bioherms (Richter et al. 1979, Kershaw & Guo 2003, 2006, Portman et al. 2005, Andrews et al. 2007). The modern shoreline contains abundant evidence of Holocene uplift in the form of raised marine notches. You can even swim out and experience the modern notch-forming environment at several locations during this itinerary.

Key objectives

- To consider the evidence for regional uplift during the late Pleistocene and Holocene; also to assess the complications wrought by minor faulting.
- To see well preserved late Quaternary marine carbonate sediments, particularly spectacular and enigmatic microbial bioherms, and to interpret these in the context of changing sea level over the past 200 000 years.

- To demonstrate the importance of robust chronologies in interpreting late Quaternary environmental changes.

Establish a base at Loutraki. As with Chapter 3, the first set of odometer readings are taken from zero at Loutraki town fountains on the north end of the main street. This itinerary is best attempted on a weekday if visited during summer: the scenic localities at Cape Heraion and Lake Vouliagmeni become crowded with tourists at weekends, and parking can be difficult.

Itinerary

Stop 4.1 Cape Heraion: scene setting and viewpoint Follow the main street north leading past the spa buildings (0.6 km) in the direction of Perachora. The road climbs out of town in a series of broad sweeping bends with road-cuts exposing Mesozoic limestone and thrust contacts with serpentinite and chert formations. After a series of very sharp hairpins (at 5.6 km), by the Panorama taverna follow the main road to Perachora village. On approaching the village (at 8.4 km) bear left at an ornamental acropolis and gardens, then left again using a small bypass road to avoid the village centre. Turn left again and the road leads down hill following a small gorge. Good views of karst scenery developed in Mesozoic basement limestone bedrock appear among olive groves, and the first glimpses of Lake Vouliagmeni (as we shall see it is actually a lagoon) occur in the distance to the west. Bear left, ignoring side roads to Strava and Sterna (at 11.3 km). At about 13 km, views of Lake Vouliagmeni develop well and the north-east corner of the lake is reached at 14.5 km (Fig. 4.1), where the lakeside taverna on the left is an ideal lunch spot. Follow the road along the linear northern shoreline of the lake, noting on your right the limestone ridge, its steep south-facing crags part of a degraded and apparently long-dead fault scarp. At the northwest corner of the lake, the road bends left and then right by a small chapel (at 16.5 km). Continue west towards Cape Heraion, where at 18.2 km the road broadens out into a car-park with views of the cape below. As you are now close to an important archaeological site, hammers are strictly forbidden in the near vicinity.

Shady spots are at a premium in the car-park, and the local cats will quickly befriend you if you rustle foodbags. Coaches can turn here, in summer there are portable toilets too. From the parking area, follow the rough path on the left, signposted to the Heraion archaeological site. About 17 paces down from the telegraph pole that marks the start of the track, the path exposes a very small section of polished bedrock limestone, which is clearly a north–south-trending fault plane dipping 40°W. The

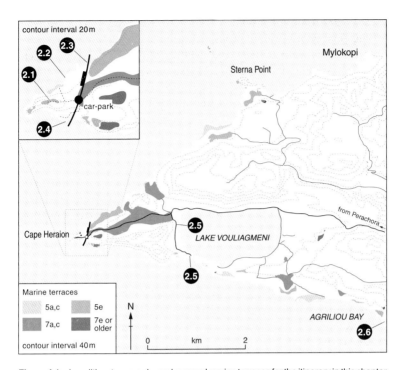

Figure 4.1   Localities, topography and mapped marine terraces for the itinerary in this chapter.

path leads down towards a small col: in the low bank on your right-hand side as you descend you see grey-weathering calcarenites that are back-tilted at about 45° towards the fault. Follow the path to the small col where the same calcarenites are more or less flat bedded (but with internal cross stratification). These relationships demonstrate over-steepening of hanging-wall strata closest to the fault, indicating that the col is downthrown *c.* 10 m relative to the car-park area (Fig. 4.1 inset; for more on this see stop 4.3). Instead of descending left to the archaeological site, continue west towards the cape and lighthouse. The path winds up from the col towards the left initially and at the first prominent right-hand bend there is a clear east–west-trending fault plane in bedrock limestone. There is no clear evidence that this fault has been active in the Holocene or late Pleistocene. Approach the lighthouse by following the path on the north side of the headland: the path suddenly doubles back and then leads up to the Meso-zoic bedrock limestone tor that forms the highest point of the cape at about 58 m altitude and where there is a deep (and unfenced) cistern cut in the bedrock limestone (38°01′45.7″N, 22°51′07.2″E). Here there are good views of the south coast of the cape, including prominent notches in the bedrock

49

just above sea level, and the much-photographed mushroom rock directly below a precipitous drop (both seen close up at stop 4.4). With regard to safety issues, beware of uncovered cisterns, vertical cliffs, hymn singers. (Coordinates: 38°01′47.0″N, 22°51′14.3″E)

The wider views are magnificent from this solitary and beautiful place, with the sounds of the sea gently lapping below and, with luck, the shriek-ing of alpine swifts as they career around the cape. On a clear day it is possible to see west to the high mountains of the Peloponnisos, south to the lumpish mass of Acrocorinth, southeast to the urban sprawl of modern Corinth and the entrance to the Corinth Canal, and northeast to the Alky-onides Islands. This is a good place to remember how the gulf is thought to have responded to sea-level changes during Quaternary glacial/ interglacial cycles (Fig. 4.2). For example, during the most recent glacial-maximum, global sea level was approximately 130 m lower than at present and the gulf was a freshwater lake whose surface would have been about 70 m lower than present sea level, held up by the Rion Sill at the western gulf entrance (Collier et al. 2000). Immediately below you to the southeast you can see the archaeological site. To the northeast you see the rocky lime-stone promontory of Cape Sterna and in the foreground a clear terrace at

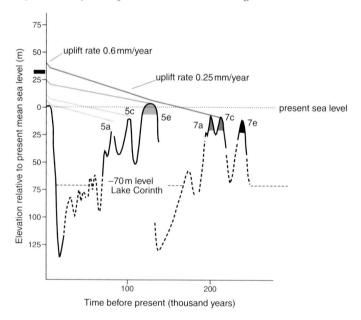

Figure 4.2 Late Quaternary sea-level curve (mainly after Imbrie et al. 1984, Chappell & Shackleton 1986, Chappell 1996) with uplift paths for dated marine terraces of the Perachora Peninsula.

Figure 4.3 Panoramic view looking east from ~58 m viewpoint of the west Perachora Peninsula (stop 4.1). The c. 25 m MIS 5e and c. 40 m 7a/c marine terrace levels are clearly indicated. The location of the pres. med 5a,c terrace level at c. 10 m and its marine fauna above the *Rivularia* bioherms (stop 4.2) is also indicated.

about 25 m elevation (Fig. 4.3). Although the original surface has been modified by agriculture, there is little doubt that it was originally a natural terrace feature (see stop 4.3). Less distinct higher terrace levels may also be seen, including the general area of the car-park at about 40 m elevation, corresponding to the MIS 7a/c terrace.

The sediments immediately beneath the 25 m terrace surface (we shall see these up closer soon) are mostly bioclastic marls and limestones with abundant marine fossils including bivalves, corals, gastropods, echinoderms and small patch reefs constructed by red algae and serpulid worms (Richter et al. 1979). Most significant at this juncture are the corals *Clado-chora caespitosa* (see Fig. 7.5), still preserved with their original aragonite skeletons. A specimen of *C. caespitosa* from this locality, dated by alpha spectrometry U/Th techniques, gave an age of 128 000 (±3000) years (Vita-Finzi 1993), consistent with coral biomineralization during MIS 5e. It is reasoned therefore that this 25 m terrace corresponds to a 25–30 m palaeoshoreline that formed during the highstand of MIS 5e, broadly the highstand of the most recent interglacial. We know from data collected in stable crustal areas elsewhere in the world that global sea level during the MIS 5e highstand was up to + 6 m relative to modern sea level; yet, here at Cape Heraion, the mapped inner edge of the terrace (i.e. the palaeoshoreline) is at about 28–30 m elevation. This demonstrates 22–24 m of uplift in about 125 000 years, a mean uplift rate of about 0.2 mm a year. Moreover, this terrace and its inner edge can be mapped at similar heights across wide areas of the Perachora Peninsula and Corinth Isthmus, suggesting relatively uniform uplift of this part of the crust, albeit somewhat disturbed by minor faulting at times.

Stop 4.2 Downfaulted microbial bioherms and land-slipped MIS 5e sediments
Retrace your steps from the viewpoint back to the col below the car-park. Standing on the col (38°01'46.9"N, 22°51'11.3"E, altitude 30 m) facing north you will see an intensely landslipped zone of chaotic limestone boulders among scrubby vegetation immediately below you and the car-park.

From the col, some indistinct and quite steep tracks lead down through the scrub (keep an ear and eye trained for Sardinian warblers here). Keep left along the path on the southwest edge of the land-slipped area towards the shore. Stick with the tracks and you should emerge in a relatively flat area about 10 m above the sea. An indistinct path leads southwest towards the cape and northeast along the coastal edge of the landslipped area. There are no particular safety issues in this locality. (Coordinates: 38°01'51.2"N, 22°51'10.7"E)

Pick your way about 50 m northeast along the clear track running parallel

Figure 4.4 Bioherms at stop 4.2 showing steep sides and relatively flat tops. The bioherms have been downfaulted by about 18 m from the unfaulted 25 m terrace to the east.

to the shore, until you can scramble down to the water's edge. Now look back southwest towards the cape and lighthouse. In front of you are several large dome-shape limestone beds (Fig. 4.4).

As will become clear soon, the dome-shape structures are downfaulted by about 18 m relative to *in situ* outcrops (stop 4.3); however, the whole block has slipped more or less intact, making the outcrop easily accessible. The dome-shape structures are bioherms (i.e. biologically formed mounds), up to 10 m high and 8–10 m wide, with steep, almost vertical flanks, relatively flat tops and "onion skin" concentric bedding (0.3–1 m thick) that parallels the bioherm top and sides. The bioherms are composed of a framework of layered and branching colonies of calcified (calcite) cyanobacterial (i.e. blue-green algae) tubes attributed mainly to *Rivularia haematites* (Fig. 4.5; Richter et al. 1979). Most of the large boulders in the landslipped area are derived from the bioherms, and freshly fractured surfaces allow close examination of the fabrics. You will notice that the bioherm framework is highly porous, and with a hand lens many of the larger growth cavities are seen to have walls coated by the coralline alga *Lithophyllum pustulatum*, which appear creamy white in contrast with the darker grey of the *R. haematites* framework. The bioherms contain no macrofossils; however, they are sometimes overlain by a ~0.5 m-thick coralline algal-bound bioclastic bed with prominent oysters, serpulids and other marine fauna, including rare corals. The contact with the underlying bioherm is distinct, and locally it features rounded limestone and chert pebbles, indicating an erosion surface. The following discussion is updated from Portman et al. (2005).

Attempts to date the *R. haematites* calcite framework of the bioherms

53

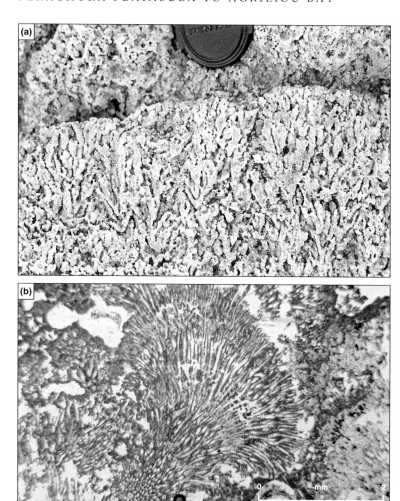

Figure 4.5 **(a)** Field photograph of bioherm fabric showing branching colonies of *R. haematites*; lens cap for scale. **(b)** Photomicrograph in plane-polarized light of a well preserved colony of *R. haematites* micrites from sample at stop 4.2. The white areas are mostly porosity with thin microspar cement rinds. The porous macrotexture of the framework is also obvious around the colony edges.

directly using U/Th techniques have failed, mainly because the calcites are contaminated with clay-mineral and organic matter (the technique works best on clean calcites). However, it has proved possible to make some corrections for this using an isochron approach, resulting in a poorly constrained age of 126 800 (+36 500/−39 300) years. Although the error is large, we can be confident that the bioherms are older than the MIS 5e high-stand sediments above them (see stop 4.1 discussion), and this means the bioherms can only be of late MIS 6 or early MIS 5e age (more on this later).

The other important field feature of the bioherms is a pale-yellow weathering, dense calcite speleothem with a smooth surface, typically lining the walls of constructional cavities (Fig. 4.6). This speleothem, known as flowstone, is up to 1.5 cm thick. It shows lobate dripstone surface textures, mostly orientated downwards relative to vertical bioherm growth. These are best seen close by in a small cave just above sea level (38°01′50.7″N, 22°51′10.6″E), where, although in a downfaulted block (stop 4.3, p. 62), they preserve their correct orientation. Do not sample these outcrops. These cements also form flat floors in the larger cavities, which together with the dripstone textures point to an origin in the vadose groundwater zone. A sample of this speleothem calcite was dated to 127 100 (+13 600/−12 200) years Portman et al. (2005) using low-resolution U-series methods. However, our recent (Andrews et al. 2007) high-resolution dates suggest an age of about 135 000 ± 2000 years, an early MIS 5e age.

Close examination of flowstone exposures reveals another level of detail. At the base of the speleothem it is possible to find a discontinuous,

Figure 4.6 Calcite flowstone coating inter-bioherm growth surfaces at stop 4.2. Note the pendant "dripstone" outer surface texture characteristic of vadose speleothems. (Two-euro coin for scale)

55

porous, botryoidal micritic crust up to 5 cm thick, which in thin section is seen to be full of green algal tubes, about 100 mm in diameter, and discontinuous stringers of the coralline alga *L. pustulatum*. At the base of this crust is a more or less continuous layer of *L. pustulatum*, typically 1–3 mm thick, which encrusts the bioherm surfaces in large growth cavities, and can form quite large pendant masses of *L. pustulatum* in the larger cavities (see stop 4.3). Search also for uncommon loose blocks containing botryoidal aragonite cements that postdate the speleothem (Richter et al. 1979).

Before leaving this area, it is also worth examining the modern seashore at sea level, where you can see a small platform or ledge of limestone, just awash at high tide, projecting from the slipped blocks of bioherms (see Fig. 4.4). This is an actively forming constructional ledge, an aggregation of vermitid gastropods (*Dendopoma* sp.) and coralline algae (*Neogoniolithon* sp. and *Lithophyllum* sp.), the encrustation often referred to as trottoir. It is ubiquitous in the Mediterranean, forming narrow ledges and shallow reefs several metres below sea level (Adey 1986). Field observations on the wider Perachora Peninsula show a planar upper surface coinciding with mean sea level, which drops steeply on its seaward side to form an overhanging submerged ledge less than 2 m thick. Because the upper surface occurs at mean sea level, trottoir ledges should be good indicators of palaeo sea level. However, palaeo-trottoir ledges are typically absent from uplifted cliffs in the area, presumably because trottoir forms below marine-notch level (stop 4.4), and is typically destroyed by progressive notch formation during uplift. Isolated patches of palaeo-trottoir are found only on the landward faces of boulders on the sheltered southern shoreline of the peninsula.

We shall return to re-examine the topmost bioclastic deposit above the bioherms at the end of stop 4.3, its significance can be appreciated only with the context of that locality in mind.

Stop 4.3 Section through the succession underlying the 25 m MIS *5e terrace level: Heraion Marl, in situ bioherms, overlying marine bioclastic sediments*
Follow the indistinct path northeast for about 100 m along the coastal edge of the landslipped area, picking your way through a jumbled mass of bioherm boulders. You should be able to see all of the features described above, and in addition you may find patches of unconsolidated marine bioclastic sediments, which, as you will see in a moment, form the sediments above the bioherms. Safety helmets are required under the cliff. (Coordinates: 38°01′52.9″N, 22°51′15.2″E)

Looking northeast you should now have a view of the cliff section shown in Figure 4.7. These cliffs are steep, with overhangs, and it is possible to

Figure 4.7 View of the cliff section comprising stop 4.3. Note the sharp erosive contact of the MIS 5e basal transgressive lag on the older Heraion marls at the bottom of the photo, the talus-like breccias of bioherm clasts above and the superb sections through *in situ* bioherms above. The well stratified 5e marine deposits may be seen to the top left of the exposure.

scramble close up to the cliff face here; however, we advise you first to observe from a safe vantage point below (binoculars help here). You should be able to distinguish several clear units in the exposure. The base comprises soft, friable, laminated, pale grey-green marls, known locally as the Heraion Marl (Richter et al. 1979). There are few fossil or sedimentary structures to be seen here, although only 300 m to the northwest the marls contain southward-dipping shelly laminae, with corals, and rhodoids of *Lithophyllum pustulatum*. The northeast dip indicates deformation before regional uplift of the peninsula, because uplift since MIS 7 has been uniform across the region: this indicates that the Heraion Marl is older than MIS 7 and, although it is widely considered to be of Pliocene age, there is no direct evidence for this and it may also be of early Pleistocene age.

The Heraion Marl is cut unconformably by a marked erosion surface at about 12 m above present sea level (Fig. 4.7). The contact is irregular and overlain by a limestone conglomerate less than 0.5 m thick, the clasts in which are mostly of calcified cyanobacterial bioherm material, sometimes bored and usually coated with the flowstone described previously. The conglomerate is overlain conformably at this locality by a locally developed chaotic cavity-rich breccia <6 m thick. As in the basal conglomerate, most of the breccia blocks are composed of calcified cyanobacterial bioherm material, and the same calcified cyanobacteria drape and bind the uppermost 2–3 m of the breccia forming the cores of the large overlying bioherms seen towards the top of the cliff (as at stop 4.2). Cavities in the breccia have some large, pendant *in situ* growths of the coralline alga *L. pustulatum*, up to 50 cm long. The flowstone (about 135 000 years old) is prominent on the surfaces of large cavities between bioherms. Although not seen clearly from this viewpoint, the bioherms are overlain by the bioclastic marls and limestones with abundant marine fossils, including corals dated to the MIS 5e highstand, as discussed at the viewpoint stop 4.1. These marine sediments infilled the irregular space left between the tops of the bioherms and then constructed the flat 25 m terrace surface viewed from stop 4.1. As the Heraion Marl and basal conglomerate beneath the bioherms were not seen at stop 4.2, it is clear that the present section is uplifted some 14 m or so relative to the sediments at stop 4.2. In fact it is in the footwall of an arcuate fault, with a maximum throw of approximately that magnitude, which runs more or less from this point, behind the landslipped area you have just traversed, under the west end of the car-park (the fault plane visible in the track as you descended to the col) and through the small embayment by the archaeological site (stop 4.4) on the south side of the peninsula (fault marked on inset to Fig. 4.1). The whole of Cape Heraion from the car-park westwards is thus in the hanging wall of this fault.

Having seen the bioherms in context with their underlying and overlying sediments, we can now reconstruct their field relations (Fig. 4.8). First of all, we know the bioherms are overlain by marine sediments of MIS 5e highstand age (stop 4.1) U/Th data from elsewhere in the world suggests that the MIS 5e highstand began at about 128 000 years ago. We also know that the bioherms are older than the groundwater-deposited flowstones that encrusted them at about 135 000 years ago. The bioherms clearly contain coralline algae, encrusting cavities within them. As coralline algae are usually fairly reliable indicators of marine conditions (indeed, the same species is helping construct the trottoir ledges today), the bioherms must have formed in shallow sea water, and that has to have been in MIS 5e (MIS 6 sea level was never high enough to flood the gulf).

As sea level rose at the start of MIS 5e, it would have cut a marked erosion surface during its rapid transgressive phase. We therefore interpret the basal erosion surface cut into the Heraion Marl and the overlying basal conglomerate as a marine transgressive surface and beach deposit, presumably formed just before about 135 000 years ago. The locally developed

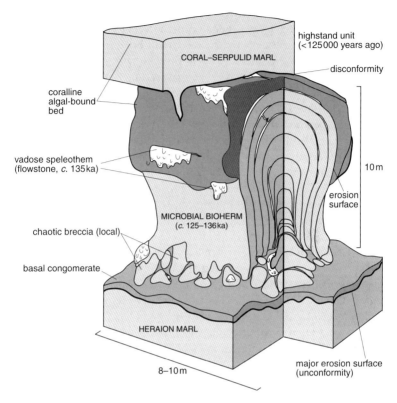

Figure 4.8 Schematic of bioherm structure showing the principal elements described in the text. Note that the coral–serpulid marl above the minor disconformity has variable thickness, probably in part determined by the size of the bioherms.

breccia above the basal conglomerate is made of blocks of bioherm material, which suggests that the growing bioherms were at least locally de-stabilized, perhaps by earthquakes associated with nearby fault movement. Bioherm growth then resumed and stable mounds grew on top of the breccia, forming massive bioherms whose upward growth tracked rising sea level. The flowstone date, which postdates bioherm growth, shows that the bioherms were in place before 135 000 years ago, so bioherm growth appears to have been localized 136 000–135 000 years ago.

The presence of the flowstone is highly significant, as it marks an episode of vadose groundwater flow through the bioherms. This can only have happened if either sea level fell or uplift occurred. In this case it is fairly easy (unusually) to identify a large eustatic (global) sea-level fall at 135 000–130 000 years ago (Thompson & Goldstein 2005), which would have caused a drop in the groundwater table, enabling vadose flowstone formation. Flowstone is not found in the overlying marine bioclastic sediments above the bioherms. Indeed, at one (unfortunately dangerous) locality nearby, the flowstone surface is clearly encrusted by oysters and serpulid worms, and, at other localities between here and Sterna, former open spaces in speleothem-lined cavities are filled by marine phreatic aragonite cements. Marine encrustation and aragonite cementation can have formed only during the highstand of MIS 5e sea level after speleothem formation, because sea levels after MIS 5e have not since been high enough to resubmerge the cements in this tectonically uplifting area (see foot of p. 62 for the one exception to this story caused by the local downfaulting already demonstrated).

Although the field observations and ages offered above for bioherm formation appear to fit together neatly, there is a snag. Cyanobacteria such as the main bioherm framework builder, *Rivularia haematites*, do not calcify in open-ocean sea water today (Riding 1982) and probably have not been able to since the late Cretaceous, when sea water globally was even more supersaturated with respect to $CaCO_3$ than it is today. Modern cyano-bacterial calcification is common only in hardwater streams and lakes. This problem led Kershaw & Guo (2003, 2006) to suggest that the Perachora bioherms grew in a freshwater glacial lowstand lake, followed by cavity encrustation by the coralline alga *L. pustulatum* during subsequent marine flooding. However, there are flaws in this interpretation. First, the bioherms should be common all around the gulf if they were indeed lake-shoreline deposits. They are not: in fact they are mainly limited to a strikingly linear outcrop (Fig. 2.1) between Cape Heraion and Sterna. Secondly, if the bioherms were of MIS 6 age, the rate of uplift required to expose them well above modern sea level today would be totally unrealistic compared to calculated rates of mean uplift between 0.2 and 0.3 mm a year. A MIS 6 age is also inconsistent with the elevation of the bioherms when compared to that of the mapped and dated MIS 5e and 7e marine shorelines.

A much simpler interpretation fits all of the observational and chronological evidence assembled above. The bioherms formed in shallow coastal waters, where fresh groundwater-sourced springs were focused upwards by a shoreline parallel fault line (Fig. 4.9), explaining the linear and isolated outcrop pattern of the bioherms. Calcification was facilitated by rapid degassing of CO_2 from calcium bicarbonate groundwater, a process that encourages $CaCO_3$ precipitation, as shown in the equation below,

$$Ca^{2+}_{(aq)} + 2HCO^-_{3(aq)} \leftrightarrow CaCO_{3(s)} + CO_{2\,(g)} + H_2O_{(l)}$$

where removal of CO_2 from the right-hand side of the equation by degassing also causes solid $CaCO_3$ to precipitate out of solution. This interpretation is consistent with detailed geochemical data from the bioherm

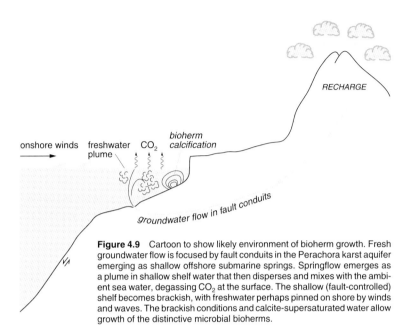

Figure 4.9 Cartoon to show likely environment of bioherm growth. Fresh groundwater flow is focused by fault conduits in the Perachora karst aquifer emerging as shallow offshore submarine springs. Springflow emerges as a plume in shallow shelf water that then disperses and mixes with the ambient sea water, degassing CO_2 at the surface. The shallow (fault-controlled) shelf becomes brackish, with freshwater perhaps pinned on shore by winds and waves. The brackish conditions and calcite-supersaturated water allow growth of the distinctive microbial bioherms.

calcites and also explains the lack of marine fossils in the bioherms, excepting the coralline alga *L. pustulatum*, which, it turns out, is tolerant of low salinities. The minor erosion surface and coralline algal-bound bioclastic bed seen at stop 4.2 record transgressive marine conditions after bioherm growth. The erosion surface with pebbles suggests that a minor sea-level fall occurred before this. As the boundstone also contains flowstone speleothem, the major sea-level fall postdated it (see Andrews et al. 2007).

You may wonder why the bioherms apparently formed only in this time period. Although there is evidence for earlier phases of calcified bioherm formation, some of which correspond to the local MIS 7a/c palaeoshoreline, the early MIS 5e ones are clearly the biggest and best. Our interpretation requires a vigorous groundwater flow, both for bioherm calcification and speleothem formation. This in turn demands abundant recharge and therefore wet climatic conditions. Interestingly, the timing implied by our dates does not obviously concur with evidence for wet climatic conditions in the wider eastern Mediterranean (e.g. sapropel events; Emeis et al. 2003) or with pollen data from lacustrine early MIS 5e sediments in northern Greece (Frogley et al. 1999). This suggests that local climatic variations in the region are important and modulate regional-scale climatic patterns (see also Tzedakis et al. 2002).

Retrace your steps back to the foot of the track descended to reach stop 4.2 at coordinates 38°01'49.6"N, 22°51'08.1"E. Pause on the top surface of the outcrops (see Fig. 4.4) here to observe an important and revealing feature developed on the top surface of the marine MIS 5e bioclastic deposit that caps the bioherms. The deposit is covered in places by oysters cemented to it, limestone cobbles and bioclastic debris, and by the characteristic borings of the marine rock-dwelling bivalve *Lithophaga*. Evidently, at some time after the MIS 5e highstand, the sediments here (but not at stop 4.3) were lithified and subsequently bored and encrusted in a littoral or shallow sublittoral marine environment. Since the younger MIS 5a–c sea-level highstands were 10–20 m lower than those of MIS 5e, it follows that the arcuate fault throwing down the western cape must have been active after MIS 5e but before MIS 5a at the latest (i.e. in the interval 120 000–80 000 years ago). The 10–15 m throw was thus sufficient here to bring the down-faulted MIS 5e sediments to a level where the 5a–c highstands were able to onlap the now lithified sediment. Fault activity ceased at some time before MIS 5a, and thereafter the whole peninsula continued its regional uplift. The rate of normal fault throw, assuming an interval of activity of about 40 000 years, was thus a maximum of some 0.25 mm a year.

Stop 4.4 Holocene marine notches and down faulted MIS 5e sediments and bioherms   Carry on back up the hill to regain the col. Do not be tempted to pick your way up the indistinct tracks that appear to lead more directly to the car-park; you would soon be lost among a bewildering array of large boulders separated by yawning gaps. Once back at the col (elevation 30 m) it is worth noting that the calcarenites below your feet are almost certainly of MIS 7 age, given that they are in the hanging wall of the fault noted above, downfaulted by about 10 m. There are other MIS 7 sediments, mainly *Rivularia* and coralline algal bioherms, just above the calcarenites here (just below car-park level), but the field relationships are complex and much confused by karst weathering. The general car-park area, with inner-edge *Lithophaga* borings at an elevation of 40 m, corresponds to the level of the MIS 7e marine terrace mapped in the area.

Descend on the south side of the col, following the clear track towards the archaeological site. Before a sharp right turn towards the chapel, a thick fault breccia associated with a prominent old (late Pleistocene, inactive) east–west-trending fault is clearly exposed on the left (see also below). The path twists west past pre-classical cisterns on the left and the modern chapel (which in season features archaeological information in a display outside) and then steeply down towards the picturesque harbour, which is over 3000 years old. Beware of spinose sea urchins and take care under the steep cliffs. (Coordinates: 38°01′43.8″N, 22°51′10.9″E)

Along the old linear east–west fault scarp on your right-hand side as you

Figure 4.10 The uplifted marine notches on the east side of Heraion harbour at stop 4.4.

reach the bottom of the path by a prominent small juniper tree, a polished and slickenslided fault plane is seen showing dip slip with minor oblique-slip displacement. The fault breccia here has cavities that contain good examples of speleothem flowstone. Although not dated here, this is pretty obviously the 135 000 years speleothem seen at stop 4.2. Walk around the western side of the small harbour and scale the hummocky limestones on the west side of the harbour entrance. You are now standing on the tops of downfaulted bioherms at elevations of about 8–11 m. As the tops of the *in situ* bioherms at stop 4.3 are at about 22–23 m elevation, we have evidence of downthrow of 11–13 m here. In fact the fault passes just west of the chapel and through the small harbour and embayment, so the limestones on the opposite (east) wall of the harbour entrance are in the footwall.

This is a good place to study marine notches. The limestones on the east wall of the harbour entrance in front of you display a clear notch at modern sea level, and at least two uplifted notches at about 0.9 and 2.1 m (Fig. 4.10). If you want to check these measurements, this is a good place to swim and wave a ranging pole about (beware of the plentiful spiny echinoids). There is also a very faint poorly preserved uplifted notch at about 3 m, seen more clearly farther along the south shore towards the lighthouse (Pirazzoli et al. 1994) and elsewhere on the peninsula. So, overall, there seem to be three persistent uplifted notches preserved on sheltered (often southern) shore-lines of the peninsula at elevations of about 0.9, 2.0 and 3.0 m, although local variations in elevation are up to 30 cm, probably attributable to contrasting exposure to wave attack and minor local Holocene faulting.

Marine notches are formed by a combination of chemical dissolution, physical abrasion and bio-erosion by borers and raspers. It is likely that chemical dissolution is a major player here, as sea water has clearly heavily karstified the modern marine limestone platforms in the surf zone all around the peninsula. It is not yet clear how surface sea water, typically supersaturated with $CaCO_3$, manages to dissolve coastal limestones, but it certainly does – all around the world.

These uplifted notches record episodes of uplift, separated by periods of crustal and sea-level stasis, and remind us that uplift occurs in short cataclysmic bursts accompanied by seismicity. We have suggested earlier that the long-term rate of mean uplift for the Perachora Peninsula is about 0.2–0.3 mm a year. This rate would imply that the upper (3 m) notch took about 10 000 years to reach its present elevation. This is tested by radiocarbon dating shell carbonate from the boring bivalve *Lithophaga* associated with the upper notch just below the lighthouse. The calibrated age from this locality was 4440–4320 BC (Pirazzoli et al. 1994), about 6390 years before present, yielding a late Holocene uplift rate of about 0.5 mm a year: the crust here has been motoring upwards recently.

Walk west from the harbour area through the archaeological site. The back (west) wall of the site has been dug into soft marly bioclastic sediments full of bivalves and the odd coral. These are the MIS 5e highstand marine sediments above the bioherms, but downfaulted here by about 10 m relative to their position just below the 25 m terrace on the northwest side of the cape. This is the best place to see these bioclastic deposits close up, although the bored and encrusted upper surface seen at the end of stop 4.3(pp. 61–62) is not present here because of a combination of natural erosion and human activities.

Using the indistinct tracks, scramble up the ridge made by these sediments and continue west until good views of the cape emerge below the lighthouse (38°01′44.5″N, 22°51′08.6″E). This is close to where Pirazzoli et al. (1994) sampled their *Lithophaga* for dating and, by the way, noted four uplifted notches, the extra one being at about +2.6 m. Kershaw & Guo (2001) suggested that the +2.05 m notch measured at the harbour correlates with the +3.2 m notch measured by Pirazzoli et al. (1994) here. This correlation requires a Holocene active fault downthrowing the harbour sequence by 1.15 m relative to the cape. A simpler solution would be to correlate the +2.05 m notch measured at the harbour with the + 1.7 m notch at the cape, putting the height difference down to differential exposure. This requires no fault and simply means the + 3 m notch is not so clear at the harbour.

Figure 4.11 The best-ever notched sea stack. The stack comprises a single bioherm, 8 m high, whose surface cap can be clearly seen.

To the southwest, close inshore you will see mushroom rock (Fig. 4.11), a spectacular sea stack eroded from a downfaulted bioherm (you can still see the distinctive rounded top) with a deeply notched base.

Retrace your route back to the car-park, but do take time to look at the archaeological site. Many of the building stones were carved from bioherm material, a relatively soft and workable limestone. Once back at the car-park it is worth walking northeast along the road for about 250 m. You soon come to a roadcut (38°01′49.0″N, 22°51′24.2″E) on the right (southeast) that exposes shelly bioclastic sands with prominent low-angle metre-scale crossbedding (to the southwest). The shells are comminuted marine bivalves (*Cardium* species and *Mytilus* species) and suggest a marine shell bank or beach face moulded by currents. Although undated, based on elevation (about 40 m) and relationships with mapped shorelines, these sediments are almost certainly of MIS 7a/c age.

Stop 4.5 West shore of Lake Vouliagmeni: palaeoshorelines, Lithophaga *borings, and uplifted beachrock* Zero your odometer and drive back towards Lake Vouliagmeni. As you approach the lakeshore (at 1.7 km) there is a road junction by a small chapel (38°01′59.1″N, 22°52′18.5″E). Turn right here and geologize on the way, eventually to arrive after about 1 km to a rough car-park (very full at weekends). The road is very narrow and not suitable for coaches, so it is best to alight here and walk, keeping a sharp lookout for cars (wear high-visibility jackets). Cars can proceed with care; there are straight sections of road where parking is safe, but pedestrians must take care because of the road traffic.

At the junction, in the bank on the north side of the road to Cape Heraion, there are patches of loosely cemented bioclastics with abundant specimens of the coral *Cladochora caespitosa* dated to 194500 (±800) years ago (i.e. MIS 7a or c) by Dia et al. (1997). From here, turn right (south) and follow the road along the south shore of Lake Vouliagmeni. After about 200 m, on a bend at the driveway entrance to a lakeshore taverna, the road bank (1.9 km at 38°01'54.5"N, 22°52'22.3"E) exposes about a metre of shelly marl with abundant *C. caespitosa*. The corals are generally very well preserved, and continuing U-series dating of specimens (unpublished work of J. Smith) from the base of these marls suggests an age of about 120000 years (i.e. MIS 5e highstand). However, notice that some corals close to the ground surface here have been dissolved, leaving just moulds. This reminds us that any dates from corals must come from well preserved original coralline aragonite. The lower part of the bank exposes a prominent 30–40 cm-thick layer of limestone pebbles above a clear erosion surface that was cut into underlying marls, with abundant serpulids and bryozoans, but also red algae and bivalves (but no corals). The erosion surface is about 8 m above sea level, but dips seawards and can be followed down to about 4 m (or less) above modern sea level in the bank of the track leading to the taverna. As the erosion surface was cut into Pleistocene marine marls some time before 120000 years ago it may represent a regressive event within MIS 5e, possibly the event 135000 years ago that allowed speleothem formation in the bioherms (stop 4.2), although it could also represent the basal erosion surface seen at Heraion (stop 4.3) if the underlying marls are of deepwater MIS 7 age. Without any constraint on the age of the marls beneath the erosion surface, it is not possible to be definitive.

Proceeding south, 200 m around the corner from the last locality the road rises over a hillock in Mesozoic limestone bedrock (2.1 km at 38°01'49.6"N, 22°52'22.6"E). On the Lake Vouliagmeni side there is a prominent limestone pinnacle and it is possible to scramble down towards the lakeshore here. On its south side the pinnacle exposes good examples of uplifted borings (10–20 mm diameter) by the bivalve *Lithophaga* (Fig. 4.12).

These borings tend to occur in bands 0.5–1.0 m wide (vertically). Living *Lihophaga* around the Perachora Peninsula are common in trottoir, just below sea level, and probably within the active notches too. With fossil examples, the uppermost borings in a band probably fix palaeo sea level reasonably well. As these examples cannot be Holocene in age (their

Figure 4.12 The *Lithophaga*-bored and notched palaeo sea stack of probable 5a and 5c age seen at stop 4.5.

elevation is too high), they are likely to have formed during MIS 5a or c. If you explore the scrubby area above the road, about 20 m south of this locality it is possible to find two clear marine notches cut into limestone bedrock at about 9 m and 12 m above modern sea level, consistent with MIS 5a and 5c ages.

Continue south along the road. Once you lose bedrock limestone there are no exposures for about 400 m as you traverse the foot of a colluvial fan. Continue south and southeast, passing a tiny lakeside chapel (2.6 km). The road ends in a large car-park on the right (2.8 km). The bedrock has now changed to serpentinized ultramafic rocks (ophiolite), which are heavily weathered. The upper surfaces are capped by a rubbly breccio-conglomerate, which is a Holocene cemented beachrock. This is seen most clearly on the seaward side of the car-park at 38°01′34.0″N, 22°52′33.8″E. The beachrock here is formed mostly from clasts of the ultramafic bedrock (up to 20 cm long axis) with some local limestone. The clasts are cemented by a mixture of micritic aragonite (dominant) and Mg-calcite cements (Richter et al. 1989), although this can be confirmed only by thin section. The clasts have often weathered more rapidly than the cement, such that the cemented rims stand proud. The beachrock can be found as seaward-dipping sheets to mean sea level, suggesting that cementation is still active, and this is confirmed by the discovery elsewhere on the peninsula of a modern coin (less than 13 years old) cemented into beachrock (Richter et al. 1989).

Beach cementation by $CaCO_3$ is favoured by a combination of microbial mediation and evaporation. The presence of aragonite beachrock cement in the Mediterranean is peculiar to areas of serpentinite outcrops, where local weathering can leach Mg^{2+} from such ultramafic rocks, which, on mixing with sea water in beach-sediment pore spaces, inhibits calcite precipitation. Elsewhere on the peninsula and in the Mediterranean generally, beachrock cementation is usually by Mg-calcite only.

As well as active beachrock formation, there is clearly beachrock here uplifted to 2 m above mean sea level. At the landward edge of the seaward-dipping beachrock surface at about 0.85 m above mean sea level, there is a prominent wave-cut notch, followed by a 0.5–1 m-high clifflet in uplifted beachrock, which makes a small terrace (Fig. 4.13). The notch was probably formed by a combination of clast decomposition (weathering) and dissolution of the cement, accompanied by limited abrasion. This + 0.9 m

Figure 4.13 Uplifted beachrock south of Lake Vouliagmeni, seaward side of car-park. Jenny Mason stands on a terrace looking at cats 1 m above present sea level. Note the irregular notch cut into the base of the higher terrace. The serpentinite basement can be seen beneath the beachrock, and the gently sloping topography is characteristic of non-limestone shorelines on the Perachora Peninsula.

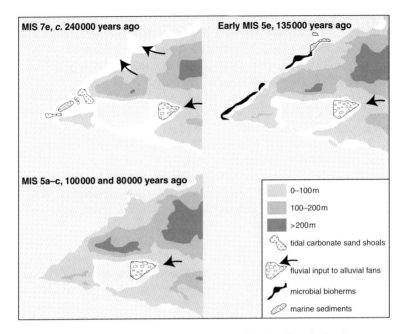

Legend:
- 0–100 m
- 100–200 m
- >200 m
- tidal carbonate sand shoals
- fluvial input to alluvial fans
- microbial bioherms
- marine sediments

Figure 4.14 Palaeogeography and palaeo-elevations of the Perachora Peninsula.

notch and terrace can be mapped fairly consistently around the Lake Vouliagmeni area and is further evidence of Holocene uplift.

Walk east along the beach until you come to the artificially cut channel that connects Lake Vouliagmeni to the sea (i.e. the lake is actually a lagoon 50 m deep). Walk along the west side of the channel back to the Lake Vouliagmeni side for a view. The huge area that Lake Vouliagmeni occupies is a giant karst depression (field karst or *polje*), now flooded by sea water. It is worth considering what this area has looked like during the past 250 000 years or so. Obviously, during glacial lowstands the area would have been simply a karst depression, presumably without a permanent lake. However, during highstands it was transformed into a marine embayment, initially connected to the gulf by a seaway north of the Heriaon Massif (Fig. 4.14). Connection to the sea was gradually closed off as progressive uplift occurred, until, by MIS 5a/c times, the palaeogeography was very similar to today, albeit with a narrow but natural connection south to the gulf, more or less where you are standing.

Lake Vouliagmeni hosts many beachside tavernas, and it is usually possible to find a shady one. Some even hire out pedalos for a more intimate look at the "blue lagoon".

Stop 4.6 Agriliou Bay: a coastal alluvial fan delta on an uplifting coastline
Drive back to Perachora, following your outbound route. At the ornamental gardens and acropolis at the junction with the Pisia–Skinos road, reset the odometer and proceed in the direction of Loutraki. At 2.6 km you negotiate the very sharp hairpins by the Panorama taverna. At the end of the subsequent straight (3.3 km), with a deep gorge to your right, there is a safe pull-in (at 38°00′19.0″N, 22°56′05.9″E) on the right-hand side of the left-hand bend, where there are fine views of the Agriliou locality below and the Perachora Peninsula to the west (Fig. 4.15). Note the form of the landscape below, with the outfall of the prominent gorge to the right, the terraced nature of the ground on the high skyline (*c.* 200 m elevation) and the clear lobe-shape bulge of the coastline with a large car-park and restaurant built just landwards (next parking place). You are looking at the surface of an alluvial fan delta of Holocene age, fed by intermittent river flow from the gorge. The boulder cluster that features at the southernmost promontory was probably deposited after a catastrophic flood event during a wetter climatic phase earlier in the Holocene. On the northwest skyline, the flat terrace is at about 200 m elevation and, at this height and below, beach gravels may be found at many localities.

Continue driving for about 0.3 km towards Loutraki, looking out for the next very sharp right-hand turning off the main route.* Once on the side road, follow it down through a series of hairpins until you reach the large restaurant car-park studded with olive trees (5.4 km). On the way down, if hiking or with a car, you may stop to examine the many new (as of 2004) exposures along the metalled road. These show deformed Jurassic mudrock, chert, serpentinite and limestone basement in complex thrust contacts defining melange associations. Resting on these basement rocks over corrasion surfaces and abutting against palaeo-stacks at elevations 150, 126, 82, 66, 54, 47, 41, 36, 29, 24, 10 m (data courtesy of Jenni Smith) are marine-beach and shoreface clastic deposits, sometimes with *in situ* bun-shape *Lithophaga*-bored clasts.

The main exposures are situated along a bluff by a small track on the west side of the car-park and along the east side. The western sequence can be studied with care for about 100 m north along the track where the basal parts are best seen; note that a house is very close to the cliff edge here, so it is not a good idea to use hammers or to attempt scaling the cliff. Safety helmets are required under the cliffs. (Coordinates: 38°00′26.1″N, 22°55′48.2″E)

---

* At 3.3 km: the turning bends back on itself and is just manoeuverable with care in a car, but not in a coach. Coach drivers will need to continue towards Loutraki until the next available place to turn and then approach from the outbound direction.

Figure 4.15 View southwest from the pull-in before stop 4.6 to show the sheer limestone cliffs bounding Panorama Gorge (right), the Agriliou Bay fan delta (centre left) and the Perachora Peninsula to the west (top left).

The bluff is the outfall of the drainage from Panorama Gorge seen at the preceding viewpoint. The first feature to note is the clear unconformity between the cream northeast-dipping Heraion Marl and younger sediments, first seen at 5 m elevation about 60 m northwards along the track, but rising landwards to >10 m about 150 m farther along the track. Detailed mapping by Jenni Smith and Clive Portman has demonstrated that this erosion surface and its overlying sediments can be traced upwards towards a terrace inner edge (at a break of slope with overlying and equivalent-age alluvial fan sediment) at about 27 m elevation. The erosion surface looks similar to that seen at stop 4.3. Here it is penetrated by burrows that have been infilled by younger sediment from above. The erosion surface is overlain by a thin limestone boulder and cobble lag, whose clasts are well rounded and sometimes imbricated, often featuring *Lithophaga* and clionid sponge borings. This is succeeded by up to 0.5 m of coarse sands and gravels, with cross stratification in places, and then by up to 2 m of poorly sorted breccio-conglomerate with both angular and rounded clasts up to boulder size. One prominent large boulder has oysters cemented to its surface. All the above units thicken seawards. The presence of coarse clastics above the unconformity suggest pulsed fluvi-odeltaic input into a shallow marine environment.

Near the corner of the bluff, at the beginning of a low wall where the southern end of the track meets the car-park, a distinctive stratigraphical unit – a bioturbated sandy marl (about 3 m thick) with abundant colonies of *Cladochora caespitosa* – lies above the breccio-conglomerates. These are 7 m above mean sea level and they weather out for easy collection. Unfortunately, these corals have yielded highly variable U-series dates ranging from 100 000 to 200 000 years, presumably because of variable post-growth diagenetic alteration. The unit is also rich in marine bivalves, including *Acanthocardia*, *Turitalla*, *Glycimeris* and *Chlamys* spp. The upper part of this marine unit contains erosion surfaces and conglomerate lenses. The overall interpretation is a shallow marine (<20 m depth) shoreface setting.

The fossiliferous marine deposits are truncated by a prominent erosion surface seen clearly above the corals at 9–11 m above mean sea level. The erosion surface is interpreted to have been cut when sea level fell. Above the erosion surface are open-framework, well sorted, fine conglomerates showing seaward-dipping cross stratification. These are best seen at the western end of the car-park along the seaward-facing bluff and are interpreted to be the beachface deposits of a wave-dominated fan-delta shoreline. The upper 2 m of conglomerate rests on another erosion surface and contains rounded and angular clasts. This deposit is interpreted to be attributable to a renewed sustained progradation of fluvial-derived sediment from the gorge into the marine littoral environment.

The southeast margin of the car-park (38°00′22.0″N, 22°55′48.2″E) has a good exposure equivalent to the upper part of the sequence described above. Above a basal limestone boulder lag (some with clumps of adnate corals) there are up to 5 m of seaward-dipping clinoforms that generally decrease in dip from 15° to zero. The clinoform beds comprise couplets of sharp-based grain-supported angular gravels and laminated low-angle cross-stratified pebbly sands. The coarser units often fine laterally down dip (seawards). The angular gravel units are often capped by a layer of bun-shape limestone clasts bored by *Lithophaga* and sponges on their upper surfaces. The couplets are interpreted to be depositional records of pulsed flood events that were then subject to periods of prolonged marine bio-erosion and sand deposition on a fan-delta beach face or upper shoreface.

Walk southwest for a hundred yards or so to the modern beach and note its delta-planform shape to the west and east. This is the marine fan-delta shoreface to the periodically active fluvial catchment issuing from Panorama Gorge. To emphasize this, continue southeast along the beach, noting the large boulders more or less at mean sea level. You are at the major fluvial outlet to an earlier Holocene fan lobe, which is being reworked by modern surf. The boulders are now being rounded on their margins and upper surfaces, and are colonized by *Lithophaga* and clionid boring sponges. They must have arrived during a very large flood event at some point earlier in the Holocene, but when? As you continue southeast around the curve of the delta shoreface, there is a basement limestone sea stack with a notch at about + 2 m, followed by a small bay and then a headland and prominent boulder field. The western side of these boulders exposes good examples of trottoir (see stop 4.2.) at modern beach level. The boulders to the east may be seen to bank up against a palaeocliff (Fig. 4.16) cut in basement limestone below a prominent notch at + 1.9 m. Below the notch are excellent examples of *Lithophaga* borings. A calibrated radiocarbon age of 6800 years was obtained from a *Lithophaga* shell at + 1.6 m elevation at this locality. This date demonstrates that the boulders arrived on the delta surface before about 7000 years ago, probably during a wet climatic phase in the early Holocene. This *Lithophaga* shell date appears quite old, however, considering that *Lithophaga* dates from the Heraion site (stop 4.4) of 6390 calibrated years BP from + 3.1 m elevation and 4390 calibrated years BP from + 2.2 m elevation. Perhaps local minor Holocene downfaulting is to blame.

Figure 4.16 Julian Andrews and Jenni Smith are marvelling at the excellent uplifted Holocene marine notch and its flanking beach gravels featured at stop 4.6.

Further reading

Following early studies in the Perachora area by Vita-Finzi & King (1985), several authors have contributed to the present state of knowledge (e.g. Vita-Finzi 1993, Pirazzoli et al. 1994, Hubert et al. 1996, Dia et al. 1997, Morewoood & Roberts 1999, Leeder et al. 2003, 2005, Portman et al. 2005). Research and controversy continues in the area; contrasting opinions as to the origin of the bioherms are by Richter et al. (1979), Kershaw & Guo (2002, 2003, 2006) and Portman et al. (2005) and Andrews et al. (2007), and on uplift and structural history by Morewoood & Roberts (1999) and Leeder et al. (2003, 2005). No-one has written about the human landscape that underpins the Perachora Peninsula better than Dilys Powell (1957).

Chapter 5

The Megara Basin

MIKE LEEDER & ROB GAWTHORPE

The Megara Basin (Fig. 5.1), trending northwest–southeast, lies between the Gerania Mountains to the southwest and the Pateras Mountains to the northeast. The altitude of the basin area generally lies below 350 m; the surrounding mountains rise to 1350 m (Gerania) and 1132 m (Pateras). We saw the serpentinite, chert and limestone rocks comprising the Gerania Mountains in Chapter 3; the Pateras Range is composed predominantly of Mesozoic limestone. The basin contains more than 1 km of exposed marls, sandstones and conglomerates of Plio-Pleistocene age. They are best seen in the narrow coastal plain and in rapidly eroding badlands, which spread for some 5 km inland from the village of Alepohori. This whole area is in the rapidly uplifting and eroding footwall to the active Psatha–East Alkyonides–Skinos faults examined in Chapter 3.

The structure of the Megara Basin is relatively simple (Fig. 5.1): beds mostly dip at shallow angles (10–15°) to the northeast, the homoclinal dip pattern exposing progressively younger strata in this direction. The basin is thus of tilted block or half-graben form, having rotated about a horizontal axis defined by the basin-bounding fault on its northeast side. In order to achieve observed rotation across such a 10 km-wide tilt block, the total vertical throw along the basin-bounding fault must be of the order of 2 km. This is of the right order, given the elevation of the Pateras footwall and the likely thickness of the basin fill. The fault line occurs along the prominent topographical gradient that defines the Pateras Range front trending west-northwest–east-southeast for about 15 km. The Pateras uplands thus represent the remnants of the Neogene footwall to the Megara Basin, their drainage catchments still supplying sediment to alluvial fans that issue from the abandoned scarp line and which long ago coalesced to define a prominent bajada whose surface is now protected by a very thick calcrete palaeosol. The Pateras fault system itself is considered to be long dead, active extension having switched over the past million years to the coastal faults bordering the southern Alkyonides Gulf, which was examined in Chapter 3.

The southwest basin margin around the hamlet of Mazi is bounded by

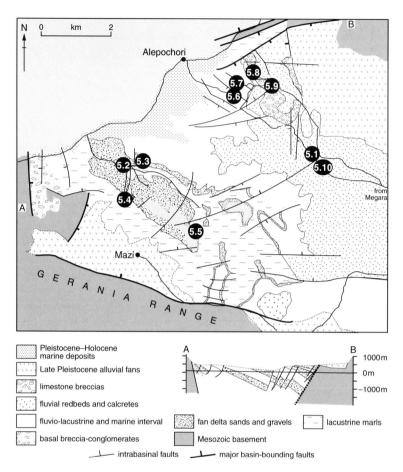

Figure 5.1 Geological map, stratigraphical subdivisions and geological cross section for the Megara Basin (after a more detailed map in Bentham et al. 1991).

a prominent northeast-dipping antithetic normal fault. In addition, many intrabasinal faults cut the Neogene basin fill. The majority of these have minor throws and are thought to represent small-scale adjustments of the accumulating sedimentary wedge during active basin extension, although some probably broke surface or caused surface depressions and ridges to form, as they acted as local controls on sand and gravel distribution.

The majority of the exposed strata are probably of Pleistocene age, although any further chronological subdivision is hampered by the lack of distinctive zonal faunas and floras. The evidence for the Pleistocene assignation is suggestive rather than definitive, based on an extrapolated age for basin abandonment of about a million years (stop 5.1) and because of the reversed nature of all palaeomagnetic samples taken from the basin fill (i.e. probably representing the Matuyama, rather than the Gilbert magnetic chron). Concerning overall basin lithostratigraphy (Fig. 5.1), units are defined as follows:

• The Paliochori Group (250 m) outcrops in the southwestern basin margin and comprises angular fluvial breccias and conglomerates succeeded by sandstones of possible tidal origin.

• The Ayio Ioannou Group (up to 550 m) comprises lacustrine marls, siltstones and sandstones, coarsening upwards into deltaic and fluviatile sandstones, pebbly sandstones and fine conglomerates with subordinate marls and silty marls.

• The Tombes Koukies Group comprises lacustrine to fluvial marls, siltstones and sandstones with the distinctive tufa limestones of the Pistarda Formation mappable over much of the basin. In the northeast part of the basin sandstones, marls and subordinate lignites crop out, with a marine interval rich in molluscan species.

• The Alepochori Group (up to 400 m thick) comprises mostly fluviatile pebbly sandstones, sandstones, calcareous sandstones and marls. The topmost units include fluvial-channel sandstones with calcic palaeosols in floodplain mudrocks and distinctive coarse limestone breccias in the Agia Sofia Formation. The backtilted southeast surface of the former basin fill is gently incised by modern drainages, with remnants of thick (up to 5 m) calcic palaeosol duricrusts on interfluves.

In terms of sedimentary environments, the major feature of successions in the central part of the basin is the presence of a coarsening-upwards sequence of marls to breccias, which records the infill of a freshwater lake system (Lake Megara) and the subsequent progradation, first of the deposits of a braided-river system, then of fines-dominated floodplains with isolated channels. Temporary marine transgressions occurred during the braided river phase, depositing richly fossiliferous marls. There is also a lateral change of facies in several mapped lithostratigraphical units, the most prominent being the early occurrence of coarse-grain sediments along the northwest basin margin and the late occurrence of very coarse-grain facies along the northeast basin margin. These lateral and vertical facies changes were controlled by the interplay of slope evolution and base-level changes brought about by the interaction of active faulting and climatically induced sea- and lake-level fluctuations.

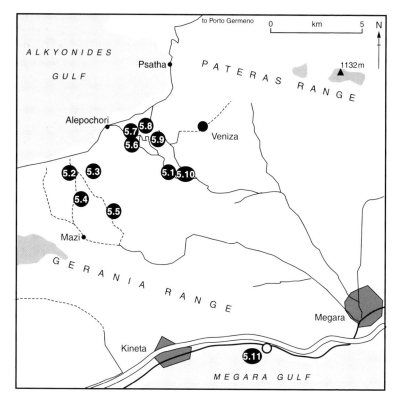

Figure 5.2 Road map to show locations.

Key objectives

- Plio-Quaternary sedimentary environments and history of a formerly active rift basin.
- Structural geology of rifting.
- Basin abandonment, uplift, tilting, erosion, drainage reversal and drainage truncation attributable to propagation of the active south Alkyonides fault system.

Travel from base to Megara (Fig. 5.2). There is very limited accommodation in Megara town. We approach from a base at Loutraki. Drive south to pick up the junction with the main Athens–Patras toll highway. The road eastwards skirts Plio-Pleistocene outcrops of the Corinth Basin, then sections through Mesozoic basement visible in many cuttings. At 18.5 km the eastern Gerania Range appears on the skyline in front of you. At 30.5 km, drive along the spectacular coastal precipice of the Skironian Cliffs, noting the many sections through Mesozoic limestones and Quaternary alluvial fans of the coastal bajada. At 31.3 km the cuttings expose the Skironian fault (see stop 5.11, p. 96) onlapped by coarse fan sediment, the latter best viewed in the layby at 33.1 km. The highway then turns landwards and cuts through the footwall to the Skironian fault, and suddenly in front appear the twin peaks of the old Megara *polis*, with the busy market town stretched beneath the east–west-trending Pateras Range, which bounds it to the north. At 41.2 km, take the Megara turnoff, following signs into town through a set of lights to arrive at a rather confusing junction with the railway line running through it at 42.8 km. Zero the odometer here.

Itinerary

Stop 5.1 Overview to observe relationships between faulting, geology and topography   At 100 m, pass the sign to the archaeological museum to the right and at 0.5 km take the right fork to arrive at a roundabout after a further 300 m. Take the signed road to Alepochori. Reach a junction over a small bridge, turn left past a supermarket, take the right at the next roundabout and head straight out of town to the northwest, avoiding all small roads leading off to left and right. We are driving approximately down the axis of the formerly active Neogene Megara Basin. On the way through the Megaran agri-landscape, with its ancient olive groves and more modern pistachio orchards, there are many opportunities to view the panorama afforded by the Pateras Mountains to the northeast. Note the strongly arcuate range front, with broad re-entrants formed at the mouths of exiting drainages. As we shall see during this itinerary, there are strong indications that the fault bounding the Pateras Range is inactive. To the southwest observe the Gerania Range. We rise in elevation along the route, the mean gradient (about 1.5°) is attributable to south-to-southeast backtilting produced by footwall uplift along the East Alkyonides–Psatha–Skinos faults examined in Chapter 3. At 14 km the road forks left at the junction with Ano Alepochori old road. Go along the straight up the hill, reaching a long right-hand bend. Taking great care with oncoming traffic, signal left and park in a layby to the left in front of a long bluff about 15 m high, which comprises the splendid section we shall return to examine as stop 5.10. High-visibility vests are recommended and beware of fast traffic on the sharp bend. (Coordinates: 38°03'45.9"N, 23°13'42.2"E)

We view west to northwest across the Alkyonides Gulf to the hanging-wall uplands of the eastern Gulf of Corinth. We are standing just below a prominent drainage divide in the form of an arcuate ridge, lying generally

Figure 5.3 Greyscale image of slopes and drainage catchments in the Megara Basin and surrounding areas. The road from Megara to Alepochori and the location of stop 5.1 are shown. The bold white lines mark the boundaries of drainage divides, with the lowest slopes represented by the darker shades. River courses are in black.

at an altitude of 300–365 m (Fig. 5.3). This separates streams draining the deeply incised ground immediately to the west, and which lead into the Alkyonides Gulf, from those drainages that descend to the Saronic Gulf down the gentler southeast slope that we have just steadily climbed. The former drainages are short and very steep in their upper courses, having incised deep gorges into the easily eroded Pleistocene succession of the Megara Basin. The landscape you are viewing is a very young one; extrapolation of footwall uplift rates determined for the active Psatha–East Alkyonides–Skinos faults (c. 0.3 mm a year; stop 3.6) gives an approximate time of a million years for the onset of fault activity and hence the abandonment of the Megara Basin. The drainage divide has resulted from cumulative southern headwards migration of drainage networks in response to the footwall uplift. The erosional products of stream incision are being reworked by a variety of sedimentary processes and redistributed around the margins and into the depths of the Alkyonides Gulf.

Stop 5.2 Coarsening-upwards sequence recording temporary infill of Lake Megara by prograding fan deltas Carry on the steep descent down the partly forested footwall badlands to the resort centre of Alepochori (Pagae in classical times, the port of embarkation for Delphi). Zero the odometer at the junction (21.7 km) with the Skinos–Psatha road. Turn left and drive 3.4 km towards Skinos. Go left at the junction signed to Agios Ioannou monastery and Prodromo, and drive 1.2 km landwards, descending the first hill to just before a drainage and track that cross the metalled road.

Park up opposite a large pine tree; coaches can turn if necessary alongside the field to the right. Walk back up the road for 40 m and turn left on the track marking the field boundary. The track turns to the right to enter a gorge below the road. Safety helmets are recommended. (Coordinates: 38°03′48.6″N, 23°09′54.4″E)

Here, splendidly exposed along the gorge wall, we have a sequence (Fig. 5.4) in the topmost Ayio Ioannou Group, which passes up from marls of the Rema Mazi Formation into sandstones and conglomerates of the Kremida Formation. Starting at the basal part of the gorge, we see marl facies often scoured clean by the periodic flash floods that roar down this drainage. The white laminated silty marls sometimes have a fauna of freshwater gastropods of the genus *Viviparus*. The laminations comprise lighter, more calcareous units a few millimetres thick, which alternate in vertical sequence with darker more silty units of the same thickness. They are persistent laterally and may occasionally show bioturbation. As noted below, the marls pass upwards into coarser interlaminated marls and siltstones. The facies is interpreted as having been deposited in the 15–30 m-deep oxygenated Lake Megara in response to seasonal variations in sediment input and $CaCO_3$ precipitation.

Above, in the walls of the gorge, we see a grossly coarsening-upwards sequence that grades from marl/silt interlaminations to very-fine to fine calcareous sandstones. The finer units include rapidly alternating 1–2 cm sharp-based silt to fine sand-grade laminae and laminated silty marls, the latter often with simple vertical burrows. The coarser laminae show occasional small symmetrical clay-draped ripple-form sets. The coarser parts of the facies include thicker sharp-based fine-to-coarse sandstone beds and thinner silty marls. The sands, 5–50 cm thick, may occasionally be pebbly or granule rich, with derived and abraded fragments of the freshwater gastropod *Viviparus*. Certain horizons show asymmetrical and symmetrical ripple-form sets, sometimes with thin lenticular clay interlaminae. Thicker sandbeds are lenticular on a lateral scale of 5–20 m. Rarely, the lenticles show flat basal contacts but with convex-up tops. Two examples have shallow scoured basal surfaces that pass laterally into

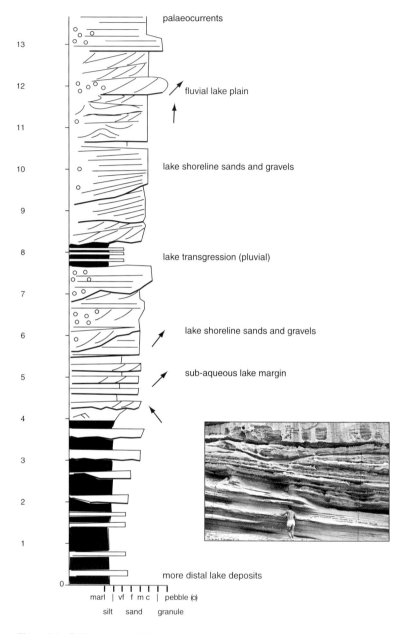

Figure 5.4 Sedimentary log with summary interpretations of stop 5.2. The inset image shows the lower part of the section to illustrate the channelized sandbeds. Vertical scale in metres.

winged lenticles. The middle-to-top parts of the coarsening-upwards facies (accessible only by abseil here) include structures indicative of slope failure and soft-sediment deformation on a variety of scales, including low-angle slide planes, listric growth faults, spectacular de-watering pipes and, clearly seen here in the lower section at the southern end of the exposure, intrusive Neptunean dykes.

The above characteristics are consistent with the section having been deposited on the shallow prograding margins of a lake subject to periodic weak wave action and to sediment underflows from distributary mouths of fluviodeltaic channels. The convex-upwards lenses of sand are interpreted as the distal portions of individual underflow lobes, whereas the winged lenses are thought to be attributable to erosive scours cut high in the mouth bars by distributary flood events. The facies is eventually erosively overlain by Gilbert-delta foreset complexes best seen at stop 5.3. The frequent evidence of sediment failure and liquefaction is indicative of slope instability, and we suspect that some of this may have been triggered by earthquakes, particularly the Neptunean de-watering structure.

Stop 5.3 Fan-delta deposits: sedimentary structures come to life Return to the parking place and follow the metalled road down hill for 40 m, and take the track leading off left up the bed of the drainage. Walk about 1 km up the track until the large Bentham's Bluff exposure is seen to the left (north). Take a short track up left to the field and view the exposure first from a distance under two pine trees. The exposure can also be overviewed on the main road that continues up hill from the parking spot towards the monastery (see travel to stop 5.5). Safety helmets are needed under the bluffs. (Coordinates: 38°03′42.8″N, 23°10′20.9″E)

The exposure (Fig. 5.5) is in the Kremida formation and it consists of sandstones, pebbly sandstones, granule and pebble conglomerates and marls. The lower coarse-grain sequence (unit 2 of Fig. 5.5) forming the brown bluff to the left rests sharply and with erosion upon marls (unit 1) of the Rema Mazi Formation. It exhibits large-scale cross-stratified sets reaching a maximum thickness of 10 m and comprises coarse pebbly-to-granule sandstone. The outcrop is a good one for appreciating the concept of apparent dips, for the faces at right angles reveal a true dip of some 20° or so to the north-northwest. Marls and sandy marls are intercalated within the coarse-grain crossbeds and these overlie them on the right-hand part of the exposure (unit 3A). A second group of such cross-stratified sandstones occurs above the marl interval at the top right-hand part of the exposure (units 3B,C), but these are accessible only by abseil. The pebble suite in the sandstones of unit 2 includes red chert (40%), ultrabasics (50%)

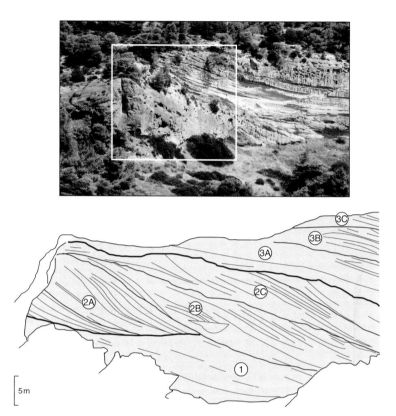

Figure 5.5 Overview and partial sketch of stop 5.3 to show the major sets of large-scale cross stratification (1–3) which represent successive progradation of lake-margin deltas into Lake Megara. The sets are broken by erosion surfaces interpreted as sequence boundaries created by the infill and drawdown of the lake waters.

and limestone (10%), sourced from the contemporary hanging-wall uplands of the Gerania Range to the southeast. Periodically the foresets are strewn with *Viviparus* shells and shell fragments. Prominent surfaces outline the overall progradational nature of the whole facies. The uppermost ends of the foresets may be terminated by erosional channels and shallow scours, which lie immediately beneath major unconformities. They range from 0.5 m-deep, 2 m-wide scours to 10 m-deep channels. Foresets are sometimes separated by thin drapes of white marl. Soft-sediment deformation is seen in the form of slump folds in foreset beds, normal faults in bottomsets and de-watering pipes in distal lacustrine sands and marls.

The deposits are interpreted as strike-parallel sections through fan deltas that periodically migrated southeast to northwest across the major hanging wall of the basin from their catchments in the Gerania Range to infill Lake Megara. Sometimes a degree of aggradation can be seen to accompany the progradation of the fan-delta foresets, indicating a gradual rise in contemporary lake level as the deltas advanced. The relative base-level changes recorded by the intercalated marls were caused either by lacustrine deepening and transgression attributable to climatic change (increased runoff or decreased evaporation, or both) or to more abrupt tectonic subsidence. Fan-delta progradation may have been aided by climatically induced lake shrinkage.

Stop 5.4 Intrabasinal listric faults   Return down the valley track to road in front of the parking place to take the track that leads south. Keep left at the fork after about 250 m, heading southeast. After a few hundred metres we pass through a deep gorge cut in facies similar to those seen at stop 5.2. After a few hundred metres more, the track leaves the valley floor and rises up the western valley side. Hereabouts the valley widens and there is an awesome panorama of the gorge cliffs to the east. Carry on to the southeast part of the cliffline where a house wall and two chimneys are visible at the top; stop here to view. A small track leads off here to the left across more open ground towards the precipitous inaccessible exposure. On the right it is possible to scramble up rough tracks on the forested gorge slopes, the *Pinus* trees here being harvested for resin. Through gaps in the trees, at about 20 m above the track, fine views can be had of the opposite cliffs (Fig. 5.6). Safety helmets are advisable if the cliffs are approached. (Coordinates: 38°03′09.5″N, 23°10′04.0″E)

The 40 m-high exposure is in the Kremida and Toumpaniari formations. Two prominent faults are visible, which shallow towards a probable plane of weakness (décollement) located perhaps only a few tens of metres below the ground surface at an horizon probably within the mechanically weak marls of the Rema Mazi Formation. The evidence for this is the form of the dip fan that has developed in the hanging-wall sediments of the fault to the right, with bedding dips steadily increasing to vertical, in rollover fashion. The likelihood that the two faults were active structures during deposition of the sediment is confirmed by strata in the hanging wall to the fault on the left; clear stratal-thickness increases may be seen as individual beds are traced towards the fault plane.

Stop 5.5 Overviews of basin stratigraphy, more fan-delta exposures and the eroded footwall cirque   Walk back to the parking place and drive east up hill in the direction of the monastery. After a few hundred metres there are fine overviews of stop 5.3. Pass the monastery. After 8.5 km scenic views of the Gerania Range develop, marked here by the line of a major antithetic

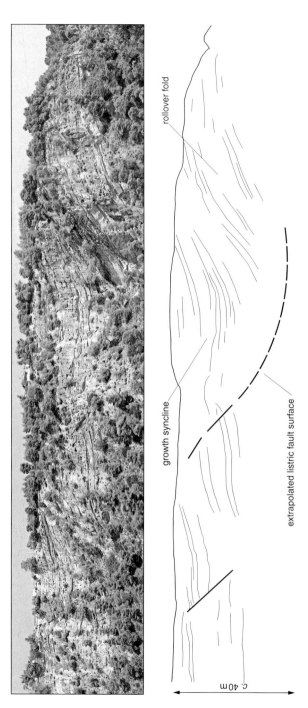

rollover fold

growth syncline

extrapolated listric fault surface

c. 40 m

Figure 5.6 Overview and partly interpreted sketch of intrabasinal listric faults developed in Kremida Gorge, stop 5.4.

fault with characteristic truncated flat-iron spurs (triangular facets) along its steep limestone footwall. At 10 km, at the apex of a sharp bend, we arrive at a charming wayside shrine dedicated to Saint Barbara.

A coach can be reversed, with care, into the wide field entrance here, ready for return to Alepochori. Walk into the field where two tracks lead off down slope. Take care on road bend. (Coordinates: 38°02′10.6″N, 23°11′52.9″E)

Here at elevation 362 m we have good views of Megara Basin setting and stratigraphy (Fig. 5.7). On the distant skyline to the northeast are the Pateras Mountains. On the nearer skyline is the arcuate trace of the 360 m drainage divide we stood below at stop 5.1. Below are steep gorges, with cliffs revealing huge exposures in Rema Mazi Formation marls deposited in Lake Megara overlain by Kremida Formation fan-delta deposits. Exactly due north, in a cliff face about 2 km distant, a fine exposure in fan-delta fore-sets may be seen with marls underneath. In outcrops to the northeast, the brown-weathering clastics of the Louba Formation may be distinguished, with the prominent ridge some 4 km distant to be viewed from stop 5.7.

Stop 5.6 Lake Megara marginal facies of marls, lignites and reedbed tufas
Return back down hill to the junction with the coast road. Turn right and drive back into Alepochori centre, turning right on the Megara road up the hill out of the village and resetting the odometer to zero.

Park the bus and hike to the excellent confectionary shop (ice cream, baklava, etc.) and petrol station on the left at 0.8 km. Cars can park more safely by the roadside or on small pull-ins. Walk up hill for 300 m to Harbour Ridges bluff, a prominent outcrop that borders the road on the right-hand side. Beware of the road traffic. (Coordinates: 38°04′58.2″N, 23°11′40.2″E)

The prominent bluff exposes an interesting sedimentary succession (Fig. 5.8) capped by a thick and intriguing limestone that may be mapped throughout the central part of the Megara Basin as the Pistarda Formation of the Tombes Koukies Group. Below are marls and thin siltstones with discontinuous darker lignitic horizons, sometimes with traces of roots. A fine conjugate pair of normal faults cuts the succession, examples of the pervasive minor extensional faulting that characterizes most exposures of the basin fill (see also stop 5.9).

The capping unit is a freshwater limestone known as tufa. Tufas are formed by the spontaneous precipitation of $CaCO_3$ from spring waters at ambient temperature. When calcium bicarbonate spring waters emerge at the land surface, they de-gas CO_2, a process that encourages $CaCO_3$

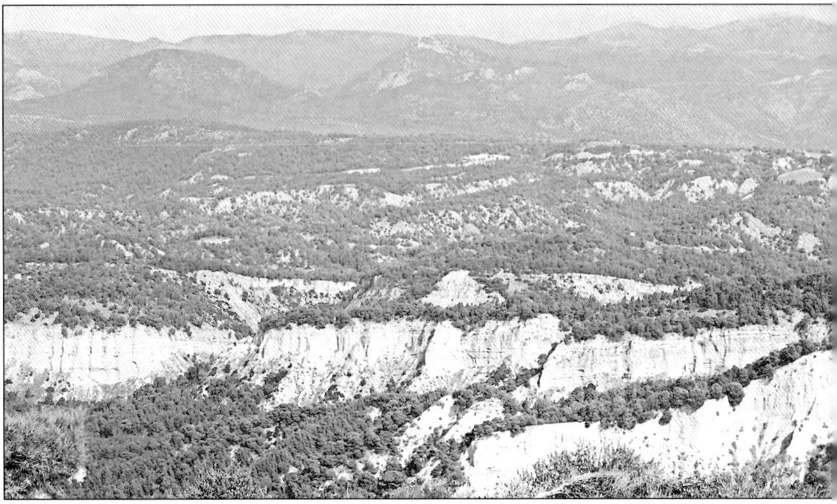

Figure 5.7 (above and opposite) View of dissected terrane taken from stop 5.5 below the erosive cirque drainage divide that borders the northeast Megara Basin.

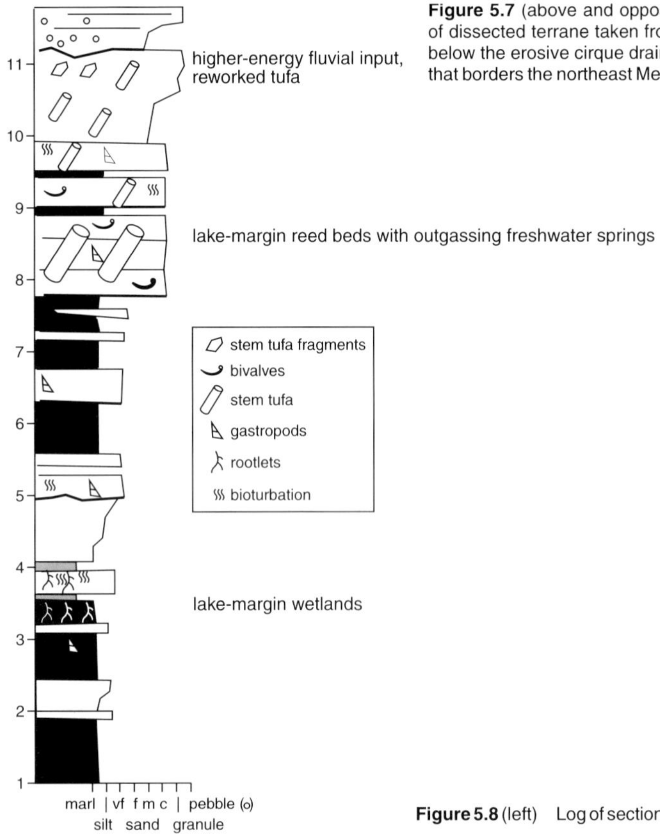
11
10
9
8
7
6
5
4
3
2
1
higher-energy fluvial input, reworked tufa
lake-margin reed beds with outgassing freshwater springs
stem tufa fragments
bivalves
stem tufa
gastropods
rootlets
bioturbation
lake-margin wetlands
marl | vf f m c | pebble (o)
silt sand granule

Figure 5.8 (left) Log of section at stop 5.6.

precipitation, as shown by the forward reaction (i.e. left to right) described by the equation on p. 60. Most tufas appear as spectacular formations at waterfalls and barrages, where CO_2 de-gasses vigorously because of water turbulence, but what we see here is different. Most of these tufa calcites have either formed as encrustations formed around plant stems (Fig. 5.9) or they appear as micrites, either detrital micrites produced by the break

Figure 5.9 Hand specimen photo of reedbed tufa located at top of section at stop 5.6.

up of the encrustations, or possibly as direct precipitates (whitings). You should also be able to find good specimens of freshwater gastropods. These types of tufas form today in marshy valley bottoms and are named "paludal tufas". Most of the encrustation occurs in very shallow water (typically less than 10 cm) although small shallow pools may develop in places. Some of the reed-stem encrustations are clearly still vertical, or sub-vertical, more or less in the position they formed, whereas in other units the encrustations have fallen over and have probably been transported a little. It is very likely that microbial biofilms play a key role in providing sites for calcite nucleation, and some of the encrustations have pustular surfaces, typical of microbial stromatolites. Notice that these tufas are also interbedded in places with lignite seams, consistent with the marshy environmental interpretation. The major (and unique for this basin) spring event recorded by the tufa is problematic in terms of origin, but it is possible that tectonically induced artesian outflow from fault-line springs could have occurred. A modern analogue in Greece is provided by the famous springs at Thermopylae, where Leonidas and his 300 Spartans bathed before the battle.

Stop 5.7 Marine incursion into Lake Megara and panoramic view Carry on hiking up hill for 60 m or so to a cutting off the road to the left. Cars can be parked by the roadside, but it is difficult at peak periods. Beware of road traffic. (Coordinates: 38°04′54.1″N, 23°12′02.3″E)

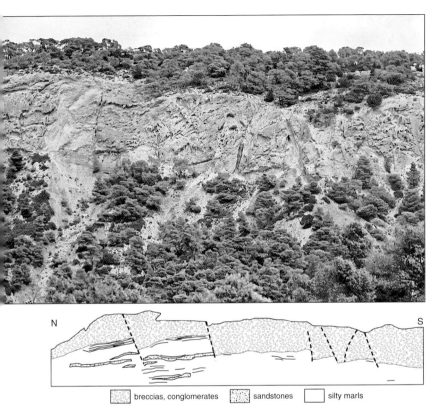

Figure 5.10 Photographic panorama and interpreted sketch of the ridge of stop 5.8, seen from the viewpoint of stop 5.7. The inset shows the detail of the northernmost listric fault; note the foolhardy author for scale. The main cliff is about 60 m high.

The exposure displays a 6 m section near the top of the Harbour Ridges Formation of the Tombes Koukies Group. It is noteworthy for its record of marly lake-margin marshland deposits with a prominent gleyed soil profile, rootlets and an overlying lignite bed. These freshwater-to-terrestrial deposits are overlain by a prominent creamy marl yielding an abundant marine molluscan fauna (*Cardium* spp., *Turritella* spp.). These represent the first undoubted evidence for marine conditions in this eastern part of palaeo Gulf of Corinth.

From the upslope termination of the outcrop, the stop also affords excellent views to the east and north. The main exposures in the crags to the right are in the conglomerates, sandstones and mudrocks of the 60–350 m-thick Louba and Agia Sofia formations of the Alepochori Group (Fig. 5.10). The hanging-wall deposits in the main cliff show an impressive group of down-to-the-west normal faults cutting the fluvial deposits. The

91

faults are listric: they have characteristic curved fault planes that decrease in dip downwards. Some show thickening of the deposits towards the fault planes, indicative of growth faulting; that is, fault displacement occurred at the same time as sedimentation. To the left of the crags, the Louba Formation dips east (rollover geometry) into the prominent syn-depositional Agia Sofia intrabasinal fault, which throws them down several hundred metres against distinctive pale Rema Mazi Formation marls.

Stop 5.8 Access for main cliff and section in lignite   Carry on walking or driving up the hill to a large sweeping bend (at 2.3 km) where the junction with the old Megara road leads off east below a large crag. Park at junction. Hard hats are recommended. Dangerous slopes on descent. (Coordinates: 38°05′01.4″N, 23°12′24.1″E)

At the foot of the cliff, immediately by the hairpin bend, a 20 cm-thick lignite in the topmost Harbour Ridges Formation is exposed. The cliff face with the listric faults viewed from stop 5.7 is accessible to the brave hearted, but survival depends upon extreme care taken on descent over loose gravel surfaces.

Stop 5.9 Sedimentary record of major bedrock failure   Hike from the parking place up hill on the old road for 900 m to a large roadcut exposure in white breccio-conglomerate (also accessible by car). Cars can be parked at the bend up hill from the outcrop. Hard hats are recommended. (Coordinates: 38°05′02.1″N, 23°12′53.3″E)

We stop first to examine a roadcut through the spectacular 10 m-thick limestone breccio-conglomerate comprising the Agia Sofia Formation and which is restricted to this northern margin of the basin fill. The deposit is sharp based, sheet like, ungraded, structureless and mainly clast supported (the matrix is a calcitic micrite cement), features consistent with a gravity-flow run-out event that may have originated as a collapse from the formerly much steeper footwall cliffs to the Pateras Range front to the north.

Farther up the road, around the hairpin bend, reddish-brown sandstones and fine conglomerates of the Louba Formation are exposed in the roadcut exposure. These overlie the Agia Sofia deposit, and the contrast between that footwall-derived deposit and this axial-drainage fluvial deposit is marked. Note the lens-shape channelized geometry of the coarser gravels and their localized and small-scale control by the many small normal faults visible in the exposures. The gravels are interpreted as the deposits of shallow stream channels. The pebble clasts include subrounded limestone, chert and serpentinite formations, all derived from the southern hanging-wall uplands to the Megara Basin, the rivers having flowed axially (i.e. structure-parallel) down the basin towards the palaeo Lake Corinth in the general area of what is now the Alkyonides Gulf. The finer horizons show blocky peds (soil structures) and calcite nodules indicative of immature calcisols forming in a semi-arid climate on the contemporary floodplains. The mudrocks are rich in the Mg-rich swelling-clay minerals palygorskite and sepiolite.

5.10 Fluvial channels and a major pedogenic duricrust   Walk back down to the transport and proceed up the steep incline back to stop 5.1 visited earlier. After parking up, assemble on the outside of the bend. For parking advice, see stop 5.1. Take great care with traffic; high-visibility vests are advisable, especially at weekends when much of wealthier Athens rapidly decamps to Alepochorian villas. (Coordinates: 38°03′45.9″N, 23°13′42.2″E)

Two bluffs expose the Louba Formation to perfection. The same general features occur as at stop 5.9, with the longer exposures here revealing the stacked sheet-like gravels (up to 25 m wide) deposited in the Louba fluvial channels. The upper bluff shows particularly impressive lateral-accretion cross stratification in the channel deposits to the bottom left of the section (Fig. 5.11).

This is also the time and place for a full appreciation that the backtilting, uplift and erosion of the Megara Basin infill has caused a major drainage reversal (as noted at stop 5.3 and stop 5.9), for these Pleistocene sediments were deposited by northwest-flowing axial drainages sourced from the Gerania Mountains to the west of present-day Megara town.

Walk some 50 m on the road back towards Megara and just down hill from the crest of the road, in the outcrops surrounding the entrance to a house, examine the super-mature calcisol duricrust exposed there. In the upper part of this sequence, over a thickness of about 5 m, the gravel units of braided channels show progressive development of $CaCO_3$ nodules. These nodules are examples of $CaCO_3$ forming in the soil zone and are known as pedogenic calcrete. Calcrete usually forms in soils and near surface sediments where annual rainfall is less than 1000 mm – typically semi-arid climates with seasonal soil moisture deficit. Ca^{2+} supply is in many cases from aeolian dust, which in this case could have either a local or far-travelled source (Saharan?). Soluble Ca^{2+} from the dust is washed downwards into soil profiles during wetter seasons. This Ca^{2+} supply will eventually cause $CaCO_3$ precipitation, as shown by the forward reaction (i.e. left to right) described by the equation on p. 60. Addition of Ca^{2+} will favour the forward reaction, whereas removal of CO_2 from the right-hand side of the equation by evapotranspiration, evaporation or de-gassing will also help.

Figure 5.11   A view of the fluviatile channel gravels and floodplain fines with palaeosols of the Louba Formation at stop 5.10.

The nodules (up to 8 cm long axis) first appear quite dispersed in the gravel matrix and these are classified (in terms of maturity) as stage 2 calcretes (stage 1 being immature, stage 5 mature). These nodules probably took many thousands of years to form, although we note that some modern pedogenic nodules seen in the road banks between this locality and Megara town are of centimetre size, so nodule growth might have been rather fast in this part of the world.

Towards the upper part of the exposure, particularly where the private driveway connects with the road, nodules become more abundant and then coalesce to form a hardpan or petrocalcic horizon. In places you should be able to spot horizons of pinkish laminated calcrete up to about 4 cm thick (individual laminae being typically about 1 mm thick), which form continuous sheets above the coalesced nodules. These laminar calcretes are of stage 4 maturity and they form where downward water percolation has been prevented (or plugged) by a hardpan horizon. Water then ponds on the plugged horizon, and calcrete forms in laminae associated with plant roots that seek out the perched water table. Fabrics diagnostic of calcified plant roots and root mats are seen in thin sections from this locality. Stable-isotope data from these laminar calcretes (courtesy of Alex Brasier) are clearly consistent with a meteoric water source and, significantly, do not suggest that evaporation has been a major process in laminar calcrete formation (evaporation concentrates the heavy isotope, so the values are typically much less negative). The $\delta^{13}C$ values confirm a contribution from isotopically negative biogenic soil carbon, probably

mixed with carbon derived from dissolving bedrock limestone clasts.

These types of mature calcrete have been interpreted to take hundreds of thousands, or even millions of years, to form. In this case a few million years is likely to be an upper estimate based on known (albeit imprecise) estimates of uplift rates (Leeder et al. 2005). Alex Brasier is currently trying to date these laminar calcretes using U/Pb dating techniques. The palaeosol here must date from initial abandonment and incision of the Megara Basin because of footwall uplift on the South Alkyonides faults. Many other exposures of the duricrust cap the low hills and interfluves that define the southeast-dipping backtilted surface, down which we now drive back towards Megara.

5.11 Two impressive but inactive fault line scarps: Pateras Range and Skironian cliffs   As we drive, note the impressive footwall scarp of the northwest margin to the Pateras Range (Fig. 5.12), the steepness of the range front slopes here, triangular facets excavated in Mesozoic limestone bedrock, and traces of debris-flow channels and levees on the steep alluvial cones that drain down the bedrock slopes. The steep range front occurs because this part of the range is actively uplifting in the footwall of the Psatha fault. This structure also cuts across the older drainages of the Pateras Range and beheads them. Farther southwest towards Megara, the Pateras Range front is noticeably less steep and more sinuous, and the outflowing alluvium of the large fans has massively onlapped and covered any traces of the old range-front faults. The fans here are not incised because of the development of the supermature calcrete noted at stop 5.10.

Drive back through Megara town to the Athens–Patras Highway and

Figure 5.12 Panorama of the Pateras Range front taken from the road back to Megara from stop 5.10. Note the diminishing steepness and increased sinuosity of the range front to the right (southeast) as footwall uplift from the active Psatha fault declines.

head west. Take the Kineta interchange and drive under the highway to arrive at the Old National Road. Zero the odometer. Drive back east in the shadow of the Skironian Cliffs, passing the Sun hotel at 3.1 km. Soon the road twists and turns through cuttings in the alluvial fans that make up the bajada. At 7.3 km arrive at a spectacular viewpoint. Parking is possible with care in small laybys. Take great care with the sparse passing traffic.

We see the faulted Skironian Cliffs and sections through the coalesced fans that make up the bajada. These are best viewed along the Old National Road on the way back west (Fig. 5.13). The coastal cliffs are also accessible at beach level and reveal the details of internal alluvial-fan growth and

Figure 5.13 Panorama of the Skironian cliffs to show the multiple sections through the Skironian fault (fault plane defines lower foreground limestone scarps) and its alluvial fan bajada afforded by the modern highway, the railroad, the Old National Road (foreground) and coastal cliff sections.

96

sedimentation by streamflow processes. Detailed mapping by G. H. Mack reveals that the fans clearly onlap the bounding fault plane along the upper bajada and are not cut by it, proof that the faults bordering this western margin to the Gerania Range are recently inactive.

Further reading

Details of the stratigraphy, structure and sedimentology of the Megara Basin may be found in Theodoropoulos (1968) and Bentham et al. (1991). The story of the drainage evolution of the area and its relation to tectonic development is told by Leeder et al. (1991, 1998, 2002, 2005), Morewood & Roberts (2002) and Goldsworthy & Jackson (2001).

Chapter 6

The Corinth Isthmus

RICHARD COLLIER & MIKE LEEDER

Throughout history the Corinth Isthmus has held a key role in the political, military, cultural and economic life of Greece. This narrow strip of land, rising to about 90 m elevation, is the natural bridge between Sterea Hellas to the north and Peloponnisos to the south. The profound cultural differences that existed across the isthmus during pre-Roman times, from the dominance of Myceneans, Athenians and Spartans, led to the prominent Mesozoic limestone inlier comprising the hill of Acrocorinth being fortified from earliest recorded times. An important settlement growing up at its foot was later to become the mighty city of Corinth, famous throughout the classical world for its pottery. It remained so until it was completely destroyed by the Romans in their revenge for impudent assertions of independence from it and other cities in the Athenian League revolts of 150–140 BC (Carthage was the other chosen target among Rome's rivals at this time). The medieval castle on Acrocorinth attests to the continuing political and economic significance of the site; it was built and controlled successively by the Frankish knights, Venetians and, later, Ottomans.

Past work on the geology of the isthmus is surprisingly sparse. The first significant description of the geology of the area was a treatise by von Freyberg (1973). Collier (1990) subsequently re-evaluated the depositional environments represented in the upper part of the canal section as being overwhelmingly characterized by a wave-dominated shoreline. This shoreline migrated in response to relative sea-level fluctuations. The key finding of Collier's study was based upon U-series dating of *Cladocora caespitosa* corals, which placed the age of each marine sequence within a late Quaternary interglacial highstand. The relative changes in sea level that led to the repeated marine sequences, separated by sub-aerial unconformities, was attributed to these glacio-eustatic changes superimposed upon a background of steady tectonic uplift. McMurray & Gawthorpe (2000) later used the Corinth Canal section and the marine terraces to the west to illustrate how variations in the rate of sediment supply influence depositional geometries on a coastal margin undergoing tectonic uplift (forced regression, in their terminology). Other contributions to the origin, geometry and deposits of the Corinth terraces include Keraudren & Sorel 1987, a pioneering wider study of the Corinthian terraces in the context of glacio-eustatic sea-level change), Collier & Thompson (1991), Collier et al. (1992) and Armijo et al. (1996). Readers may also wish to refer to Collier & Dart (1991) for a description of the Lower Pliocene early-rift sedimentary succession in the northern part of the Corinth Isthmus.

The isthmus and contiguous areas to the north and south define the Corinth Basin (Fig. 6.1), a partly uplifted structure bounded by inactive faults trending east–west, the Loutraki system to the north and the Kenchreai system to the south. The basin thus has an overall graben form. An intense gravity low in the northern part of the isthmus suggests that the sediments within the basin are several kilometres thick (King 1998). Syn-rift deposits more than 800 m thick, associated with an active Pliocene basin depocentre, occur east of the isthmus (Collier & Dart 1991). As discussed in Chapter 4, the isthmus is part of a larger uplifting area including the whole Perachora Peninsula, the Bay of Corinth, and the remainder of the modern active rift flank extending far westwards. Evidence for this uplift comes in the form of marine-terrace flights along the southern basin margin from New Corinth to Xylocastro and beyond.

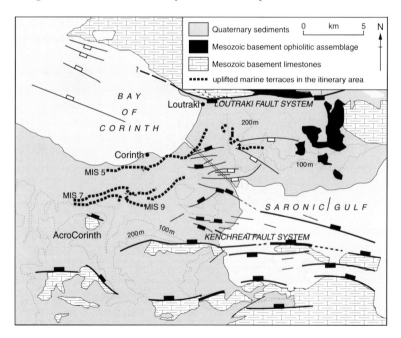

Figure 6.1 Geological structure map to define the Corinth Isthmus basin as an uplifting terrane between the major but inactive Loutraki and Kenchreai normal fault systems. Mainly after Collier & Thompson (1991), with additional offshore data interpreted by R. Collier.

Key objectives

- To view the effects of sedimentation and intrabasinal faulting in an uplifting terrane.
- To deduce the role of late-Quaternary eustatic sea-level changes in the development of coastal sedimentary cycles.
- To assess the significance of marine carbonate sediments in determining the existence of marine connections and enhanced tidal currents across the isthmus during late Quaternary times.

Itinerary

Stop 6.1 Overview of the Corinth Canal   If travelling west from Athens, pass the Elefsis, Megara, Kineta, and Agios Theodori interchanges before taking the one signposted to Loutraki. The slip road merges with the Old National Road, which leads directly to the late nineteenth-century steel bridge over the canal. If leaving from a base established at Loutraki, zero the odometer at the Loutraki fountains. Take the main road south from Loutraki to Corinth. After two broad bends, the road then turns to a southeast straight proceeding into a rather complex junction with the Old National Road at 6.5 km. Take the fork right (west) towards the old bridge (Fig. 6.2).

There is ample parking space off the road in front of a couple of

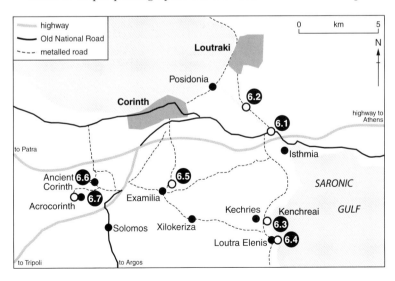

Figure 6.2   Stop maps and roads for the itinerary in this chapter.

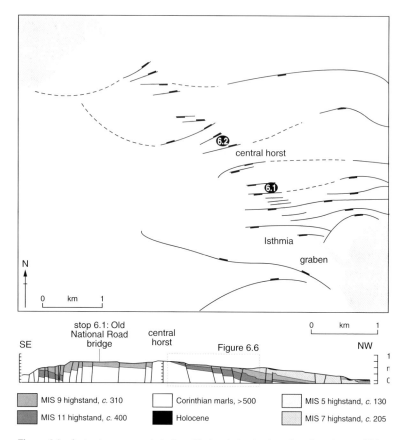

Figure 6.3 A structure map and stratigraphical and structural section of southwest cliff (as viewed from northeast bank) of the Corinth Canal (after Collier 1990).

souvenir shops just a few metres from the beginning of the right-hand branch of the bridge span. Ensure that great care is taken while crossing the bridge; zebra-crossing etiquette is almost unknown here. The bridge spans are often thronged with sightseers. (Coordinates: 37°55′43.1″N, 22°59′42.0″E)

The canal, constructed 1882–93, cuts through the isthmus, revealing an extraordinary panorama of faulted stratigraphy stretching northwest–southeast for about 6 km (Fig. 6.3). Exposures are perfect, rising from sea level at both extremities to about 90 m elevation southeast of the centre. As we look northwest, this high ground defines a prominent ridge in cream marls (Fig. 6.4) along the canal in the middle distance (Figs 6.3–6.5); this is the central horst. The Corinthian marls may be seen to be sharply and

Figure 6.4 The view southeast from stop 6.1 on the Old National Road bridge to show the south-throwing fault array, with a fine listric fault in the right foreground.

erosively overlain by up to 40 m of sandstones, conglomerates, marls and rarer oolitic limestones. All the normal faults seen cutting the stratigraphy on both sides of the bridge are throwing southeast (i.e. away from this central horst). As we shall see at stop 6.2, the corresponding faults on the north side of the central horst throw to the northwest. The symmetry of the structural architecture is thus impressive. As we shall see in subsequent stops, the uplifting isthmus developed from a structural high that was accentuated during uplift by these many small faults (throw 5–20 m), yet the faults were of insufficient magnitude to affect the overall uplift. Some of the faults seen from both sides of the bridge show curved, convex-down, shapes in sectional view that defines them as listric features; they probably extend only to shallow depths within the marls in the subsurface (see analogous features at stop 5.4, p. 85). Stratal thickening within some sedimentary wedges defined by the faults indicates that they were active during deposition.

Stop 6.2 Central eastern wall of the canal   Drive back on the Loutraki road, along the straight, and at the first left-hand bend indicate left; as the bend begins to turn right, turn left, crossing the offside lane into a bypassed metalled road behind the back of a builder's yard.

There is ample parking space in the bypassed road section. Hard hats

Figure 6.5 The view southeast to the central horst, taken from the beginning of the walk towards stop 6.2 to show the north-throwing fault array. See the panel of Figure 6.6 for architectural details of the exposures. The Old National Road bridge of viewpoint stop 6.1 can be seen on the far skyline.

are needed under the sometimes loose conglomerate outcrops. Extreme care is needed along the canal cuttings. Do not go more than a metre or so from the edge. The paths were those originally constructed for excavation of the canal, so they have been in place for more than a hundred years and may be considered safe. (Many wild tortoises live in the undergrowth about here; they graze after rain.) (Coordinates for entry point: 37°56′23.9″N, 22°58′49.4″E)

Walk southeast through a rubbish tip towards a prominent bridge carrying a pipeline across the canal. Immediately before the access steps to the bridge, turn left down to the broad track that eventually runs parallel and adjacent to the canal bank. Walk southeast for about 300 m until the track narrows and you see white Corinthian Marl outcropping in the cutting to your left. Carry on until you reach a prominent fault, whose trace can be extrapolated southwest across the canal to a similarly orientated exhumed fault plane. Telegraph pole 1 (Fig. 6.6) occurs just to the southeast.

The normal fault you are standing in front of has an impressive damage zone comprising many subparallel shear planes. It has a throw to the north

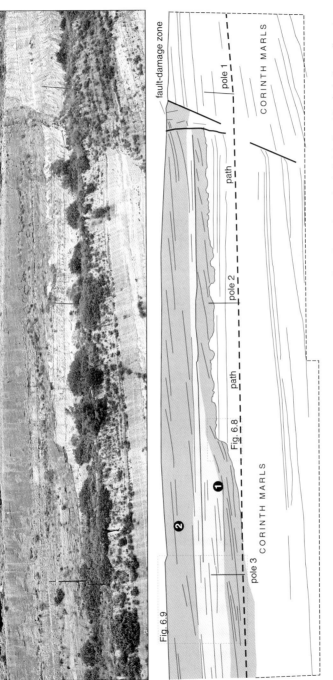

Figure 6.6 Panorama and architectural panel of the northwest bank exposures of stop 6.2 taken from the southwest bank. Points 1 and 2 refer to the sedimentary sequences defined in the text. The irregular line marking the top of the Corinth marl is the karstified and calcreted calcreted lowstand surface also discussed in the text and featured in Figure 6.7.

of some 15 m or so, but the sediments in the hanging-wall section on the opposite bank shows no evidence for growth during deposition (cf. stop 5.4). It juxtaposes cross-stratified conglomerates against the marl. Walk back northwest in the hanging wall of the fault for about 50 m, noting as you do so the sharp and irregular junction between the marls and the conglomerate; this is a major stratigraphical boundary of regional significance. Stop at a place roughly halfway between telegraph poles 2 and 3 (Fig. 6.6), where the marls suddenly descend a metre or so to the northwest. Here, various salient and revealing features may be seen.

First, examine the marls. In many ways the planar laminations between silt and muddy silt in the lowermost few metres resemble those seen in the lacustrine marls deposited in Lake Megara (stop 5.2) and a *Viviparus* freshwater fauna is common locally. Across the opposite cliff, sedimentary slumps and small slide planes may be seen lower down in the Corinthian Marl sequence. About 4 m below the irregular upper surface of the marls is a prominent sharp-based bed (*c.* 0.15 m-thick) comprising marine algal rhodoliths about a centimetre in diameter. After a further interval of barren marls, a spectacular unit rich in marine fauna, including *Thalassinoides* bioturbation, large scallops and superb *in situ* colonies of the coral *Cladochora caespitosa* may be seen. In view of what we see subsequently in the stratigraphy, we are evidently looking at the earliest establishment of fully marine conditions across the palaeo Gulf of Corinth. The corals here have dates that exceed limits to the U-series method (i.e. more than 400 000 years). The shallow marine environment represented by this unit was short lived, for the top surface of the marls is obviously well cemented, sharply defined and intricately arranged to define pedestal-like shapes up to 0.4 m high (Fig. 6.6, 6.7). This irregular surface is thought to result from emergence and karst dissolution during regression, the pedestals essentially similar to clints or flachkarren.

Overlying the karstified marl surface are 8–10 m of conglomerates, which are overlain in turn by 5–10 m of sands (Fig. 6.6). At our viewing position we see that the sudden downturn in the marl/conglomerate contact noted above may be matched across the opposite bank in the footwall of the small fault just examined. It is evident that the stepdowns are palaeocliffs; the overlying conglomerates bank up against them and build out from them as cross sets, clearly seen against the 2 m-high palaeocliff on our side of the canal (Fig. 6.8). The conglomerates also overlie the palaeocliffs and extend onto the flanks of the central horst. They comprise well rounded, often discoidal, clasts of serpentinite, limestone and chert with rare fragmented *Ostrea* shells. Internally the unit shows characteristic low-angle (10–15°) cross stratification that consistently dips northwest (Fig. 6.9). Many of the depositional wedges defined by the cross sets show

Figure 6.7 Close-up view of sharp, irregular contact between karstified Corinthian marl and overlying beachface conglomerates of sequence 1.

Figure 6.8 Close-up view of the palaeocliff (*c.* 2 m high) cut into the karstified Corinthian marl, together with offlapping beachface conglomerates.

Figure 6.9 View of clastic sedimentary deposits of sequences 1 and 2 as seen at the position of the third telegraph pole (Fig. 6.6). The base of sequence 2 beachface conglomerates overlying shoreface sands of sequence 1 occurs at approximately the top of the pole.

down-dip fining, often accompanied by a decrease in the dip angle of the foresets. The conglomerates show all the characteristics of beachface deposits, recording a seaward-prograding beach, off shore to the northwest in this case. They are identical in texture, composition, grainsize and structure to the modern prograding beaches that extend from Loutraki to the northwest mouth of the canal, and may logically be deduced to have had the same provenance, that is, from the large catchments that drain the southern flanks of the Gerania Mountains to the north, via longshore drift. The huge compositional and grainsize contrast with the underlying marls evidently records a major change in sediment dynamics in the Corinth Basin and adjacent areas. We shall return later to speculate on the possible significance of this in terms of tectonics and also on the sea-level implications of the unit.

Carry on walking northwest about 75 m or so, past splendid exposures in the beachface conglomerates at telegraph pole 3 (Fig. 6.9) to a subsidiary cliff face. Here the overlying well sorted medium to coarse sands are clearly exposed, revealing fine internal wave oscillation and combined flow ripples, their exquisite internal herringbone and chevron cross laminations often draped by silt lenses. These are succeeded by longer-wavelength hummocky cross-stratified surfaces. Overall, the diminution of grainsize and the pronounced fairweather and storm-produced structures indicate that a deepening event occurred to drown the previous beachface environment; one envisages an overall shoreface environment, perhaps down to 10–20 m water depth. Above the sands there is a markedly erosive junction with low-angle dipping conglomeratic cross sets of almost identical appearance (a little coarser) to the older unit just observed. On the west bank of the canal, this second conglomerate also occupies the top of the cliff and, as it is traced laterally, it descends the cliff face to reveal further palaeocliffs cut into the underlying sands and conglomerates. In outcrops to the northwest this clastic unit fines laterally until it becomes a thick marly unit that has yielded specimens of *Cladochora caespitose,* dated by U-series to about 312 000 years (MIS highstand 9). A third bipartite beachface conglomerate and shoreface sand clastic unit erosively overlies this second example in exposures about 1 km or so farther northwest. Corals collected from stratigraphically equivalent deposits to the southeast of the central horst gave U-series ages of about 205 000 years (MIS highstand 7e).

This is a good time and place to bring together the overall stratigraphical significance of the sediments we have seen to date. The Corinthian Marls contain evidence for the first truly marine transgression into Lake Corinth at some time prior to about 400 000 years ago. Sea level subsequently fell, leading to exposure, soil formation and karst development. Both this karstic erosion surface and that overlying the succeeding bipartite clastic unit are overlain by beachface deposits. The clastic deposits between the marl and the overlying disconformity erosion surface define stratigraphical sequence 1. Three further sequences (2, 3, 4) may be viewed in the 2 km-long continuous exposure to the northwest, the latter being marked by a marine oolitic carbonate deposit. The erosion surfaces are interpreted as lowstand features, with reworking and palaeocliffs cut during subsequent marine transgression. In each case following highstand, plentiful coarse clastic beach sediment arrived to cause northwest coastline progradation, fine examples of forced regression (shoreline movement due to net sediment flux or uplift, or both). Further transgressive highstands are recorded in sequences 1 and 3. Since the highstand deposits of sequences 2 and 3 are dated to coincide with MIS 9a and 7e respectively, it is logical to assume that the periodicity of the major cyclicity is approximately 100 000 years and that this must coincide with the eccentricity-driven band of Milanković cyclicity. This would probably make the Corinthian Marl marine transgression of MIS 13 age, sequence 1 of MIS 11 age and sequence 4 of MIS 5 age. Our field evidence for multiple highstands within individual sequences, as seen in sequence 1, implies that the MIS highstands are multiple, as known from the marine isotope record (see Fig. 4.2 for example), and that in this case the magnitude of the first highstand was smaller than that of the second.

Regarding tectonic implications of the stratigraphical section, the horst-like form of the isthmus, with its successively offlapping marine sequences, requires broad and probably steady uplift over at least the past 500 000 years or so. Following the general principles outlined in Chapter 4, and making allowances for throws on minor faults, the magnitude of this uplift may be calculated as a minimum of 0.2–0.3 mm a year. This is remarkably similar to the longer-term rates deduced for the Perachora Peninsula. Upon this uplift minor displacements were superimposed on the normal faults that define the central horst. These minor faults have been likened to the movement of keys on an uplifting piano; they disturb only the surface of the keyboard, not its overall upward trajectory. The origin of the faulting is obscure, as it seems certain that individual fault structures pass horizontally into the mechanically weak Corinthian Marl substrate at no great depth (i.e. they have a shallow listric form). They imply that a modicum of extension accompanied uplift.

The influx of coarse coastal clastics at about 400 000 years ago has two possible tectonic implications. The first is that it records that uplift of the Gerania Mountains hinterland had reached a critical elevation such that the drainage catchments on its south flanks could deliver a sustained sediment flux into the Bay of Corinth for distribution by longshore drift south-westwards. This uplift was partly regional, for it is clear that the uplift affecting Perachora–Isthmia–Corinth (extending eastwards 20 km to Kineta) has occurred at the constant rate of 0.2–3 mm a year over the past 250 000 years. It must also have included a large contribution from foot-wall uplift associated with the south Alkyonides fault system discussed during stop 5.1 (p. 80). Perhaps it is just coincidence, but the values derived for the latter are very similar to the regional rate. The second implication is that the uplifting central isthmus horst and adjacent areas had reached a critical water depth by that time, such that, during high-stands, an emergent platform connection could be made to the Loutraki outlet of the major Smarpsi fan drainage, which sources the coarse sediment. Prior to this critical depth being reached, the submarine and previously sub-lacustrine isthmus was out of range of coarse clastics, and the Corinthian marl facies dominated sedimentation, possibly in water depths of 100–150 m at the onset of uplift (perhaps a million years ago).

Walk back along the canal path. At about 100 m farther, look across the canal to see a magnificent listric growth fault with clear thickening of sequence 1 in its hanging wall.

Stop 6.3 Kenchreai – submerged port to ancient Corinth   Drive back on the Corinth road to the canal. Turn right, cross the bridge and drive to the first traffic lights. Take the left lane and turn left. Pass under the Athens–Patras highway and drive for about 2 km, latterly down hill through various roadcuts through Quaternary sediments until views of the open Kechries Bay are seen ahead. At the bottom of a long downhill straight, signal left and turn into the Kenchreai archaeological site.

There is ample space in the site car-park, but may be full on summer weekends. There are no particular safety issues here. Swimming is pleasurable, as echinoids are absent from the fine gravely substrate. You can snorkel around the submerged archaeological site.

This delightful spot is very suitable for a lunch stop. The main focus of interest is the fact that the archaeological site was the Aegean port for ancient Corinth and that it is partly submerged. Walk out to the farthest line of remaining walls and view the submerged early Christian church in which Paul of Tarsus is said to have preached. Since early Christians are not known to have worshipped up to their knees in water, the inference is that subsidence has occurred in this area within historical times. Sudden subsidence of the port area is indicated by submergence of valuable unpacked building materials (Ruth Siddall pers. com. 2006). This fact is of some interest because we have seen thus far that the whole isthmus area is uplifting and had been doing so for a very long time before the spread of Christianity. Now, look to the southwest and observe the prominent linear range front of the Onia Mountains. We are standing in the hanging wall of the east–west Kenchreai fault (see Fig. 6.1), which bounds this range; one possibility is that there has been post-Pauline movement on the otherwise dead fault. A reported discovery of late-Holocene scarps along the range-front bajada seemed to solve the mystery (Noller et al. 1997, Goldsworthy & Jackson 2001), but their relation to possible archaeological field boundaries in the area (Ruth Siddall pers. com. 2006) needs further investigation.

Stop 6.4 Loutra Elenis – Helen's cold tub Exit the site and turn left to resume the drive south for about 1.5 km. About half-way through Loutra Helenis village find a small sign for a minor road on the left to the Loutra Helenis spring site. Pass through a maze of small streets, heading for the sea. There is ample space in the site car-park. There are no particular safety issues in the locality. (Coordinates: 37°52′30.1″N, 22°59′54.0″E)

This is also a fine lunch stop to view the fabled bathing spot of Helen of Troy. Walk north along the coastal path, passing alluvial fan deposits and possible uplifted beach gravels (some clasts are sub-rounded). The brackish and rather foul-tasting spring issues from the base of the cliff where limestone outcrops. After an initial submerged phase as it first issues out from Helen's pool (the beach gravels are conspicuously white here because

111

of the lack of marine algae) the water rises and issues seawards to form a buoyant surface jet.

Stop 6.5 Examilia – oolite sandbodies of tidal origin   Drive back on the road to Corinth. After 0.7 km, take the left fork signed to Xylokoriza (before reaching the Kenchraie site). There are fine views of the Onia Range to the left as you drive along. Wind your way through the sleepy village of Examilia and at 6.9 km bear right at the double junction and take the left fork with a vineyard on the left of it. At 7.8 km, buses can park. If you are in a car, take the junction on the right. At 7.9 km go into the lane past the house, the quarry is on the left. At 7.8 km, a coach can turn at junctions. Although there are no local safety issues, the quarry can be suffocatingly hot in summer. Hard hats are needed in the quarry under old faces.

Quarrying during Greek Hellenistic and Roman eras has spectacularly exposed the core of a major oolitic carbonate sandbody. This material was the preferred building stone (locally referred to as *poros*) in this area when Corinth was politically and economically prominent. Quarries along a west-southwest–east-northeast trend reveal the internal features of a carbonate sandbody some 2 km long and which was originally up to 20 m high. The feature was deposited on a flat-lying marine terrace parallel to a fault scarp to its southeast. The oolite immediately postdates coral-bearing sediments dated to about 200 000 years, suggesting an MIS 7 age for formation and deposition of the ooids (Collier & Thompson 1991).

   Vertical quarry faces parallel and perpendicular to the axis of the longitudinal dune superbly expose high-angle sets of cross stratification, which dip to the north and to the south, away from the axis of the structure. The sets, up to 15 m high, are at the angle of repose for spherical sand grains ($>25°$). Characteristic features of good sorting and up- and down-slope pinch-outs are consistent with avalanching of sand down steep slipfaces; these are seen in both north- and south-facing sets. Individual 0.02–0.1 m sets typically alternate between medium and coarse sand, with parallel laminations and fine sand or silt-grade laminations, the latter representing quieter episodes of deposition out of suspension. The fines often contain the upward limit of burrows, indicating that these reflect periods of relatively low sedimentation rates and faunal colonization. Further evidence for fluctuating or periodically reversing current conditions is provided by small-scale planar cross stratification on the north flank of the structure. These crossbeds show both westerly and easterly palaeo-current directions. Together with the enhanced bioturbation of finer-grain laminae and the apparent alternation of low and high sedimentation rates, the evidence points to a tidal origin for the bedform.

High-angle erosional discontinuities are seen along the flanks of the large-scale bedform, evidence for differential erosion and deposition along its length. This, with the steep angles of sets to north and south, leads to the linear bedform being interpreted as the product of oblique to bipolar tidal currents flowing along its length, analogous to tidal-current ridges on modern tide-dominated clastic shelves. The implication is that, during deposition, significant tidal currents in shallow warm supersaturated waters were funnelled and concentrated through a narrow seaway, across what is now the southern Corinth Isthmus (Fig. 6.10). Tidal currents were fast enough to generate the large-scale linear tidal bedforms (>1.0 m per second for sand-grade sediment), even though the Mediterranean has a limited tidal range. The envisaged situation is analogous to the tidal amplification seen in the Messina Straits between Sicily and Calabria, or around the island of Andros, where large tidal dunes are forming today in the otherwise microtidal Mediterranean.

The complete preservation of an isolated tidal dune is perhaps surprising, but clearly the feature has been exceptionally preserved immediately after emergence (because of some combination of tectonic uplift and

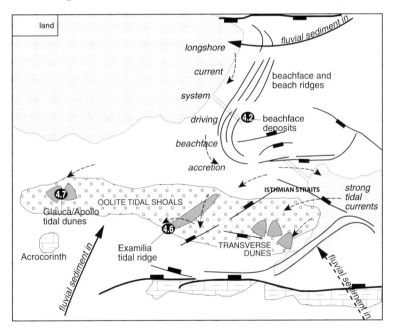

Figure 6.10 A simplified palaeogeographical reconstruction of the Corinth Basin during deposition of MIS 7 transgressive marine sequence 3, with strong tidal currents generated in the Isthmian Straits along the Isthmia graben connecting the Saronic and Corinthian gulfs (after Collier & Thompson 1991).

113

sea-level fall). As might be expected, there is evidence of widespread secondary dissolution of the oolitic carbonates internally, along joints and dissolution pipes. The top surface of the oolite is capped by an orange-weathering stage 4 laminated calcrete, up to 5 cm thick and best seen on the southern dune flank along the entrance to the quarry adjacent to the house. Remember that this calcrete formed underneath a soil profile of unknown thickness (probably more than a metre) and was not the surface crust as it appears to be today. Efforts to date this calcrete by U-series methods have failed so far because of detrital contamination

Stop 6.6 Ancient Corinth – a glorious, sad ruin set into MIS *7 marine carbonate deposits* Drive back to Examilia and find the road west to ancient Corinth. Drive through the village, following instructions for the archaeological site. After strolling through the site you can refresh yourselves at Nico's taverna, last on the right as you walk back into the main village street from the site exit on the excavated Lecheo road. Among the many souvenir shops, seek out the carver of olive wood whose small shop is 50 m out from the square at a left junction with the road you came in on. There is ample space in the site car-park. There are no local safety issues to worry about.

The archaeological site at ancient Corinth presents a wonderful opportunity to combine human history with geological history. Some of the ancient buildings either have foundations on Late Pleistocene tidal dunes or are carved directly from these oolitic carbonates. The most spectacular example is the Fountain of Glauke, which is carved from a 7.5 m-high transverse dune (Fig. 6.11). The walls of the structure provide a 3-D image of the internal architecture of this large tidal dune, with its avalanche sets dipping to the southwest. Each set is typically 5–15 cm thick (up to 30 cm) and either thickens or thins down slope, with varying intensity of bio-turbation. Bounding surfaces are planar or convex-up. Additional features, such as small-scale contortion of bedding surfaces and brittle failure, record syn-sedimentary slip on the avalanche sets. You may walk out the plan form of the Fountain of Glauke dune on the floor of the archaeological site. It shows a broadly elliptical form, with an east–west axis (Collier & Thompson 1991), and forms one of several isolated oolitic carbonate dunes in the ancient Corinth area. Another is located at the northeast corner of the Temple of Apollo. Together, they provide further evidence of tidal-current influence through the southern isthmus seaway during the late Pleistocene (Fig. 6.10).

The Temple of Apollo is constructed on oolitic sediments associated with these large dunes, and the top surface of these sediments (immediately below the temple) is coated by a laminar calcrete that is very probably,

Figure 6.11 Part of the 3-D exposure through oolitic tidal transverse dune afforded by the Fountain of Glauka at the ancient Corinth archaeological site.

although not demonstrably, the same laminar calcrete as that seen at stop 6.5. Walk to the eastern end of the low hill, upon which the temple sits, and here, among the ruins of buildings on the west side of the Lechaion road (and also on the north side of the temple hill), you can see a cross section in the sediments underlying the oolite, exposed by quarrying. Working upwards, the unconformity is clear between the Corinthian Marl and about 1.5 m of overlying conglomerate, and this is followed by about 2 m of granular sandstones and then a prominent 0.70 m-thick bed of white pedogenic calcrete nodules in a brown sandy matrix (stage 2; see stop 5.10). The oolites beneath the temple complete the sequence. The granular sands below the calcretes also contain traces of vertically elongated nodular carbonates (known as rhizocretions) that calcified around the roots of plants.

Stop 6.7 Acrocorinth: overview of Corinthian terraces, Isthmus and Gerania Mountains Leave the archaeological site car-park and take the road signed to Acrocorinth. Climb the steep hill and park at the top before the entrance to the Venetian curtain walls. Coaches can turn here and there is a friendly taverna. The best views are from the very top of the hill, through the imposing gated entrance and above the encircling curtain walls. There is ample space in the site car-park.

Basement rocks comprising Mesozoic limestones and ophiolitic facies are exposed on the road up Acrocorinth. This prominent hill probably represents the rotated and uplifted, but now part-buried, footwall, of a major intra-basinal fault. The summit affords glorious views north and west. The Gerania Mountains and Perachora Peninsula beckon from across the Bay of Corinth. The narrow isthmus is never better viewed. West of Corinth, the coastal landscape is dominated by a prominent staircase of terraces that increase in altitude westwards, the older and higher terraces reaching about 600 m south of the towns of Xylokastron and Kiaton (Keraudren & Sorel 1987, Armijo et al. 1996). Individual terraces are separated by scarps 10–30 m high and each terrace is typically 0.5–1.5 km wide. Terraces are typically overlain by a few metres of marine shoreface deposits, as exemplified in sections on the road south from Kiaton towards Souli, although this changes farther to the west towards Xylokastron, where thicker deltaic deposits overlie each terrace surface.

The morphology of the Corinth terraces has been described by various authors (Keraudren & Sorel 1987, Armijo et al. 1996, McMurray & Gawthorpe 2000) and the age of some has been correlated through U-series dating of corals with late-Pleistocene marine highstands (Collier et al. 1992). Each terrace is inferred to have been exposed in lowstand sub-aerial environments, karst formation of the underlying pre-terrace marls locally being preserved, but then cut by marine erosion and planation during a period of sea-level rise and highstand to produce the broad low-angle wave-cut surface of each terrace flat (McMurray & Gawthorpe 2000). The difference in altitude between successive terraces records the tectonic uplift that accumulated in the interval between successive marine highstands. The capping sediments record various levels of sediment influx and coastal progradation during highstand. Generally, each terrace unconformity is overlain by a transgressive basal conglomerate lag with bored pebbles and coralline red algae, and foreshore conglomerates to beach and upper shoreface conglomerates. These may then pass upwards into crossbedded upper shoreface sands and lower shoreface siltstones, analogous to the marine sequences observed in the Corinth Canal section. These upward changes to deeper water are interpreted as deposition during periods of sea-level rise; and, where coarsening upwards occurs in the upper part of a terrace depositional sequence, these are inferred to represent deposition during the highstand to early glacio-eustatic sea-level fall of one of the late Pleistocene interglacials or major interstadials (McMurray & Gawthorpe 2000).

The Corinth terraces and the offlapping marine sequences exposed in the Corinth Canal section in many ways demonstrate comparable relationships between coastal morphologies and depositional sequences developed during a period of fluctuating sea level and continuing background tectonic uplift. The difference between the canal sequences and the terraces is that, in the latter case, successive depositional sequences rarely overlap, but form discrete tabular bodies that step down towards the coastline across each terrace scarp. This reflects contrasting rates of tectonic uplift (increasing westwards towards Xylokastron and beyond) and sediment supply around the Gulf of Corinth margin. Sediment-supply rates would have been relatively high in the isthmus area, close to the source of ultra-basic conglomerates on the northern margin of the Corinth Basin. This contrasts with the area to the west of Corinth, where rates of coarse-grain sediment flux were lower, as a result of catchments being developed mainly in Neogene fine-grain formations.

Further reading

Collier (1990), Collier & Thompson (1991) and Collier et al. (1992) provide detailed accounts of most of the localities featured in this chapter. Keraudren & Sorel (1987) provide the first correct modern interpretation of Quaternary Corinthian terraces in terms of differential regional uplift accompanied by global sea-level change. The best overview of the terrace story and its tectonic background is by Armijo et al. (1996), with contrasting views on the tectonics by Leeder et al. (2003).

Chapter 7

The western gulf and the Eliki fault system

LISA McNEILL

The Eliki fault dominates the topography and geomorphology of the southern shoreline of the central-western Gulf of Corinth in the vicinity of Akrata, Diakofto and Aigion. The Aigion fault to the north is another major fault system upon which the town of Aigion sits. This itinerary will visit both fault systems and their related sedimentary deposits. Information about the offshore geology will also be provided from recent geophysical surveys (Fig. 7.1).

The Eliki fault can be divided into two segments, each about 15 km in length. The eastern segment extends from the Krathis to Kerinites rivers, where it steps right to the western segment, which continues west to the Meganites River (Koukouvelas et al. 2001, DeMartini et al. 2004, McNeill & Collier 2004; Figs 7.1, 7.2).

The Aigion fault cuts across the major fan delta formed by the Selinous River extending onto the continental shelf to the east and forming distinct segments to the west (Koukouvelas & Doutsos 1996, McNeill et al. 2005a, Palyvos et al. in press). The section of the fault from the continental shelf to the Meganites River overlaps with the western segment of the Eliki fault; both faults appear to be recently active (DeMartini et al. 2004, Palyvos et al. in press).

Key objectives

- To see evidence for tectonic uplift in the Holocene to late Quaternary by active normal fault systems.
- To see and imagine effects of earthquakes within the region.
- To appreciate the scale and internal structure of the giant Gilbert fan-delta sequences of the region.

119

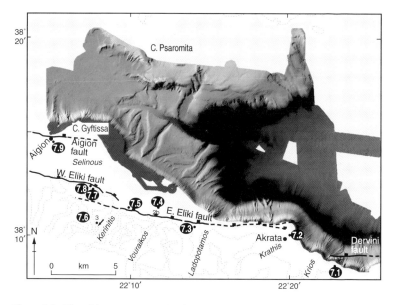

Figure 7.1 Map of the western-central gulf, showing onshore traces of major faults (including Eliki, Aigion and Derveni), approximate stop locations and offshore bathymetry from a 2003 multi-beam survey (McNeill et al. 2005a). Bathymetry ranges from narrow shelf platforms extending to 200 m depth and slope canyon systems feeding a prominent east–west trending axial submarine channel leading to the central basin plain at more than 800 m depth.

Historical earthquakes

Seismic activity in the western Gulf of Corinth is high, with many significant (magnitude >6) events in historical records (e.g. Papazachos & Papazachos 1989, Ambraseys & Jackson 1997) including several within the past few decades. The most significant earthquake (and associated tsunami) in historical times is that of 373 BC, which destroyed Helike, the principal city of ancient Achaea, and its entire population (Marinatos 1960, Soter & Katsanopoulou 1998a); the event and the remains of the city were extensively recorded in classical and subsequent literature (Pausanias in AD 174, Seneca, Strabo, Ovid). The city was thought to be located close to modern Aigion, either off shore or on the Selinous fan delta. Offshore investigations have yielded no evidence of the ancient city.

Extensive archaeological campaigns have been undertaken by Dora Katsonopoulou, Steven Soter and colleagues to identify the location and any remains of the ancient city. Their results suggest that the ancient city is probably buried between the Selinous and Kerinites rivers. Despite the historical information available, it is not known which fault produced this

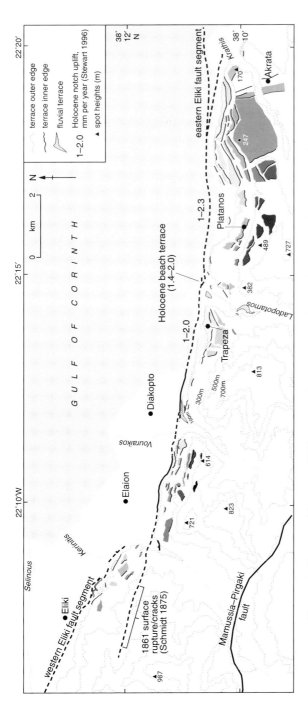

Figure 7.2 The trace of the eastern Eliki fault and marine terraces identified in its footwall block (from McNeill & Collier 2004).

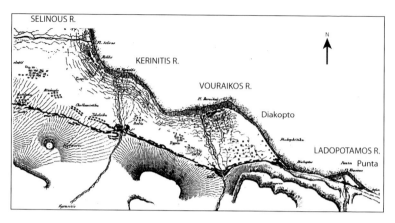

Figure 7.3 A fault rupture map from the 26 December 1861 earthquake which ruptured the eastern Eliki fault (Schmidt 1879).

earthquake. The generation of a tsunami may suggest that the fault was located off shore; however, the tsunami may have been produced by an earthquake-triggered submarine landslide. Soter (1998) argues that evidence for subsidence of the ancient city during the earthquake, and its location in the hanging wall of the western Eliki fault, suggests that this fault generated the earthquake. However, it is also possible that shaking-induced liquefaction and collapse of the coastal plain may have produced secondary subsidence.

In 1861, an earthquake ruptured the eastern segment of the Eliki fault between the Ladopotamos and Kerinites rivers, and continued westwards into the foothills west of the Kerinites (Fig. 7.3). Eastward offshore extension of the rupture is unknown. The estimated magnitude is 6.6, based on damage reports and probable rupture length (Ambraseys & Jackson 1997). Schmidt (1879) mapped the surface rupture – one of the first studies of its kind – and noted evidence of extensive liquefaction throughout the fault hanging wall, including sand volcanoes (see Fig. 7.11, stop 7.5). Cracks up to 2 m wide and 1–2 m of subsidence were reported, but it is unclear which, if any, of these features are a result of primary displacement on the fault. Possible rare fault scarps from this historic event may still be observed (Koukouvelas et al. 2001); however, McNeill & Collier (2004) suggest that the geomorphological picture is complicated by human intervention, removing and creating scarp-like features in the form of agricultural terracing.

More recently, earthquakes in 1992 and 1995 occurred in the western gulf (e.g. Bernard et al. 1997). Both epicentres were located on the north shore of the gulf, but neither have been confidently assigned to a fault

122

source. Initial reports of the 15 June 1995 Aigion earthquake (magnitude 6.2) indicated that little displacement (less than 4 cm vertical) occurred on the Aigion fault (Koukouvelas 1998), but most scientists disregard this fault as the source of the earthquake based on its epicentral location (e.g. Bernard et al. 1997) and suspect that Aigion fault movement is secondary. Bernard et al. (1997) favour a low-angle (33°) north-dipping fault plane source, which would be located in the offshore gulf. Possible fault candidates are identified in seismic profiles (e.g. Stefatos et al. 2002, McNeill et al. 2005a), but these faults are all steeply dipping (> 50°), at least in shallower basin sediments.

Geomorphology and palaeoseismology

The geomorphology of the gulf coast is strongly influenced by the active fault traces described within this guide. In the western Gulf, the uplifted footwall blocks of the Aigion, Eliki and Derveni faults produce topographical highs of up to 1000 m, with older uplifted fault blocks at higher elevations to the south. In contrast, the modern subsiding hanging wall is dominated by fan-delta systems of the coastal plain, which prograde into the gulf. The fans are fed by fluvial systems that incise into the uplifted footwall blocks of ancient fan-deltaic sediments (Dart et al. 1994, Malartre et al. 2004). This results from the active fault jumping northwards with time; therefore a previously subsiding hanging-wall region currently uplifts in the active fault footwall block. The incision process produces impressive river gorges such as the Vouraikos south of Diakofto, through which a narrow-gauge railway passes to the mountain town of Kalavrita. The river gorges provide cross sections where the topset, foreset and bottomset sequences of the Gilbert fan-delta systems are beautifully preserved and easily visible. Within this itinerary, a stop within the uplifted Kerinitis Gorge and fan-delta sequence is included.

The location of the fault traces close to sea level has produced plentiful evidence of uplift within the active fault footwall block. Sea-level markers such as eroded notches in basement limestone and wavecut platforms have been preserved and uplifted during the Holocene and Pleistocene (Fig. 7.2; e.g. Armijo et al. 1996, Stewart 1996, McNeill & Collier 2004). Many of these features are associated with datable faunal material, such as bivalves and corals, which can be dated using techniques such as radiocarbon (for material less than 50 000 years old) and U-series. As we have seen in preceding itineraries, these materials allow tectonic uplift to be quantified and the slip rates of faults to be estimated.

Palaeoseismological trenching of the Eliki and Aigion fault traces has been carried out by several groups (e.g. Koukouvelas et al. 2001, Pantosti et al. 2004, McNeill et al. 2005b). By exposing the youngest sedimentary layers displaced across a fault, they can be dated (often using radiocarbon dating of charcoal) and displacement measured. Results in this region provide some indication of frequency of earthquakes and minimum surface displacement during events. Accuracy of displacement per event and fault slip rates from these techniques are complicated by multiple surface fault splays and correlation of surface slip with slip at depth. These problems also affect earthquake recurrence accuracy.

Geodesy and rates of extension

Geodetic surveys, including reoccupation of triangulation sites following a 100-year period and GPS campaigns, have provided an indication of current extension rates across the gulf (e.g. Davies et al. 1997, Clarke et al. 1997, Briole et al. 2000, Avallone et al. 2004). Results from these campaigns suggest increasing extension from the eastern to the western end of the gulf. The western-central gulf described within this itinerary represents the part of the rift that is extending at the highest rates, up to 10–15 mm a year.

Offshore fault systems

Recent marine geophysical surveys have provided greater insight into the geometry of offshore fault systems, the offshore extensions of major onshore faults, the influence of major faults on sedimentary systems within the basin, and the regional rift formation and evolution. Stefatos et al. (2002) and McNeill et al. (2005a) have focused on the region off shore from the sites described within this itinerary. Both datasets include seismic reflection profiles imaging the basin sediments of the gulf; the latter authors also conducted a seafloor bathymetric survey (Fig. 7.1). The seafloor morphology is dominated by the submarine canyons fed by southern-shore river systems, which in turn feed into a major axial channel. An uplifted bathymetric high in the northern gulf is controlled by normal faults, which contribute significantly to extension of the gulf. Therefore, this part of the gulf is deformed by distributed extension on several major faults, both on shore and off shore, each contributing to the high net extension rate (10–15 mm a year). In general, offshore datasets indicate that many significant faults are located off shore and a complete understanding of Corinth rift deformation requires examination of both the onshore and offshore structure.

The Corinth Rift Laboratory

The Corinth Rift Laboratory was a large-scale project funded by the European Union to undertake drilling and monitoring activities of the Aigion fault system to investigate fault mechanics, fluid flow and earthquakes. The project drilled to a depth of 1000 m, intersecting the Aigion fault at 750 m depth. Sensors have been installed in various locations for long-term monitoring of strain, seismicity, fluid pressure and chemistry, including some sensors in the boreholes. A special volume documenting some of the results should be consulted for additional information (summarized by Cornet et al. 2004).

Logistics

All distances are recorded along the Old National Road (ONR) working east and west of the New National Road (NNR) junction with the Old National Road within Paralia Akrata (Fig. 7.4). The itinerary can be run from east or west, the sequence here being from east to west. Several optional stops are also suggested for those who have more time in the area, including good locations for lunch and a tourist excursion on a narrow-gauge railway. These stops are mostly suitable for small parties only – pay attention to notes on road access and parking issues. Coach parties can easily do stops 7.1, 7.2, 7.6 and 7.8. In addition, groups of small to medium size in a few cars can do stops 7.5, 7.7 and 7.9. If plenty of time is available, add in lunch and viewpoint 7.3 and excursion 7.4. (N.B. stop 7.7 is for a single small group only.)

Stop 7.1 Egira: limestone notches uplifted by the Derveni fault   At 3.7–3.8 km east from ONR/NNR junction, the site is located at the 60 km roadsign and

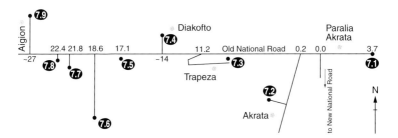

Figure 7.4 A generalized map of stops along the Old National Road (ONR), not to scale. Distances in kilometres from the junction between the Old and New National Roads in Paralia Akrata. Field trip stops 7.1–7.9 in bold.

just after a right-hand bend, just to the east of a left-hand turn to Limoni (port). There is parking for a coach or 2–3 cars on the north side of road at a new house plot (as of June 2005). Walk across the house plot to climb down over the cliff or follow the path to the beach on the east side of the foundations and small headland. Watch out for snakes in the under-growth if walking down to beach. Pay attention to traffic. (Coordinates: 38°8′34″N, 22°22′46″E)

This site lies in the uplifting footwall of the Derveni fault (see Fig. 7.1). In locations around the Gulf of Corinth where Mesozoic limestone basement is exposed, notches eroded at mean sea level are often preserved and uplifted, and marine organisms can be identified above sea level. Here, a small outcrop of brecciated limestone is draped by fossiliferous marine deposits, which include the shallow-water coral *Cladocora caespitosa* (Fig. 7.5), tidal-dwelling vermetid gastropod *Dendropoma petrauem*, and the boring bivalve *Lithophaga lithophaga* (Fig. 7.6; Mouyaris et al. 1992, Stewart 1996, Stewart & Vita-Finzi 1996, Pirazzoli et al. 2005). These are uplifted several metres above sea level and many have been radiocarbon dated by Iain Stewart and colleagues. The dates and elevations of the marine fauna have been used to determine how fast the fault footwall is uplifting rela-tive to sea level. Most of the dated fauna live close to mean sea level or within a known depth. The maximum elevation of *L. lithophaga* and their borings (Fig. 7.7) represent approximate mean sea level, whereas *C. caes-pitosa* in the wider Mediterranean live 5–25 m below sea level.

As also discussed in Chapters 3 and 4, erosional notches form by bio-logical, physical and chemical erosion at mean sea level. Notches are com-monly associated with fauna such as *L. lithophaga* borings (Pirazzoli et al. 1994). The ideal environment for deep notch formation is where sea level and the land remain at a constant relative position for some time. The notch is preserved if sea level falls, the land rises, or both occur. At this locality, Stewart (1996) records erosional benches and notches at about 1.8, 3.0, 3.5, 5.2 and 6.8 m above sea level. Pirazzoli et al. (2005) identify marine fauna up to an elevation of 9.2 m, dated at about 7200 calibrated radio-carbon years BP.

The results of dating and faunal elevations at Egira reveal uplift rates of 0.8–2.5 mm a year over the past 9000 years (Stewart 1996). Pirazzoli et al. (2005) calculate slightly higher uplift rates of more than 2.5 mm a year. Footwall uplift would be one component of slip on the Derveni fault relative to sea level, with hanging-wall subsidence increasing total vertical displacement. There is currently some debate about whether the uplifted erosional features and associated fauna represent co-seismic (earthquake-induced) uplift or gradual uplift through time. Stewart argues that the

Figure 7.5 An uplifted *Cladocora caespitosa* coral in life position at Egira site; compass clinometer for scale (length 10 cm).

data at Egira and other locations along the Eliki fault to the west indicate pulses of enhanced tectonic activity during the Holocene and that individual notches and fauna can be linked to specific earthquakes.

Stop 7.2 Eastern tip of the eastern Eliki fault, Akrata At 0.2 km west of the ONR/NNR junction, take the left-hand turn to the south to Akrata. At 2.7 km along this road, take a right-hand turn just after a school playground. Take the left-hand fork along this road and then the right fork at the church, and stop at 0.5 km from the school turnoff. A small parking area is available on the left side of the road, with space for several cars or one coach here or on the road. Farther along this road from the parking area, a gravel track continues to the north as the paved road takes a right turn. Take the gravel track and walk to the valley overlook near a bench and children's playground. (Coordinates: 38°9′34″N, 22°19′9″E)

This viewpoint is located at the edge of the Krathis Gorge on one of the late Pleistocene marine-terrace surfaces uplifted in the eastern Eliki fault footwall block (McNeill & Collier 2004; Fig. 7.8). Estimation of late Pleistocene terrace age is determined from dated coral samples, comparison with uplifted Holocene deposits and correlation with highstands of the global sea-level curve (McNeill & Collier 2004). The terraces at this elevation (200 m) are thought to be stage 7 in age (about 200 000 years; McNeill &

127

Figure 7.6 *Lithophaga* borings in brecciated limestone. Fossil bivalve shells remain in several of the borings; this material can potentially be radiocarbon dated. Compass clinometer for scale (length 10 cm).

Collier 2005), giving an uplift rate of about 1 mm a year. The terraces are superimposed on uplifted Gilbert fan-delta sediments deposited by the palaeo Krathis River. The stratigraphy of the fan-delta sediments can easily be observed from this viewpoint. Typical Gilbert fan deltas are commonly coarse grained and often formed in synrift settings. They were originally documented by G. K. Gilbert in Lake Bonneville, Utah, USA. They comprise three main components: subhorizontally bedded topsets originating as sub-aerial fluviatile to shallow marine deposits; steeply dipping foresets of coarse sediment deposited by gravity flows; and subhorizontal bottomsets formed at and beyond the base of slope. The detailed stratigraphical architecture of these components is strongly influenced by fluctuations in sea level. The understanding of how such systems evolve over time, the geometry of potential reservoir bodies and how their formation is influenced by rift faulting, is naturally of great interest to the hydrocarbon industry.

At the top of the cliff at this site, topsets continue laterally into seawards-dipping foreset deposits (Fig. 7.9). Subhorizontal bottomsets are

Figure 7.7 *Lithophaga* borings on the cliff face at the Egira site. Notice that many borings exist from sea level to 2.5 m and are prominent again at the top of the cliff (5–6 m). The person is approximately 1.6 m tall.

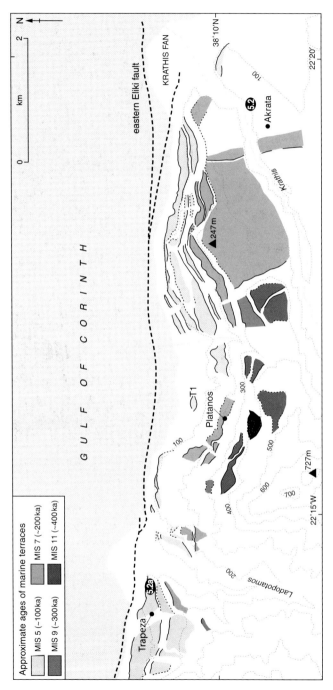

Figure 7.8 Marine terrace surfaces of eastern tip of eastern Eliki fault (after McNeill & Collier 2004).

Figure 7.9 The Krathis Gorge, revealing topsets and foresets of uplifted fan-delta deposits. The eroded cliff surface at 200 m is a late Pleistocene wavecut platform formed during a sea-level highstand within oxygen isotope stage 7 (200 000 years ago). View towards the northwest.

more rarely seen as they are often buried at the base of the sequence. Channel deposits can also be identified in the cliff stratigraphy truncating surrounding foresets. The terrace surfaces (200 000 years old) have been displaced by approximately north–south-trending normal faults in this area, forming a graben. These faults can also be seen on the other side of the gorge to the northwest (Fig. 7.8). The faults are thought to be related to complex deformation and warping within the step-over zone between the Eliki and Derveni faults.

This location is ideal for sketching the components of a typical Gilbert fan-delta sequence, as is stop 7.6 in the Kerinitis River valley.

Stop 7.3 Overview from Trapeza terraces (suitable lunch location)   At 11.2 km west of the ONR/NNR junction take the left-hand turn to Trapeza. The road hairpins up to the top of a terrace. After 1 km, take a left turn to a parallel road; bear right and park along this road for overview. The suggested stop is 0.4 km along this road, or continue to end of road and park near the overlook to east. Park at the side of the road, where there is room for several cars or a coach. Walk to terrace edge for an overview of the Eliki fault system and the Gulf of Corinth. For smaller parties there are several good tavernas here for a lunch or drink stop at the terrace edge. (Coordinates: 38°10'34"N, 22°14'1"E)

From this excellent vantage point in the centre of the fault, the entire eastern Eliki fault system can be observed (Fig. 7.8). The village of Trapeza sits on one of the late Pleistocene marine terraces uplifted in the fault footwall. Correlation of terrace elevations with highstands within global sea-level curves by McNeill & Collier (2004; see also Armijo et al. 1996 for Corinth terraces) suggests that the terraces here were formed during MIS 5 between 125 000 and 80 000 years ago, with a late Pleistocene uplift rate of about 1–1.5 mm a year. These authors estimated a slip rate of 4–7 mm a year for the eastern Eliki fault, based on these uplift rates (taking into account subsidence in the hanging wall).

Immediately below the terraces at Trapeza, the small fan delta of the Ladopotamos River progrades into the gulf in the fault hanging-wall block. Notice that the main roads and railway line all sit close to the Eliki fault trace at the base of slope. Uplifted beach deposits are preserved on this fan delta, dated at 4200–3500 years ago and uplifting at 1.5–2 mm a year (McNeill & Collier 2004). To the west of Trapeza, the major Vouraikos fan delta at Diakofto and the extensive Kerinitis (see stop 7.6) and Selinous fan deltas beyond at Aigion can be seen. To the east, the rocky uplifting coastline of the eastern half of the eastern Eliki fault towards Akrata can be seen. Several sites along this section of uplifting coast were studied by Stewart (1996), Stewart & Vita-Finzi (1996) and Pirazzoli et al. (2005). They dated uplifted fauna and recorded erosional notch elevations to determine Holocene uplift rates for the Eliki fault (see also stop 7.1).

On the north side of the gulf, the coastline is thought to be subsiding. However, recent studies of offshore fault patterns (see Fig. 7.1; Stefatos et al. 2002, McNeill et al. 2005a) and of the coastal geomorphology suggest this may be an oversimplification. The dramatic topography to the north is a result of basement uplifted during the Hellenide collisional phase.

Stop 7.4 Rack and pinion railway, Vouraikos Gorge, Diakofto   At 14 km west of the ONR/NNR junction, take a right turn into Diakofto. Follow the road into the town centre at the railway line. The station serves the mainline trains (from Corinth, Athens and Patras) and the rack and pinion railway. The station and railway are signposted from the ONR. The area is busy and parking difficult, but there are normally a few spaces for cars; a short walk back to the station may be required.

The narrow-gauge railway winds its way up the Vouraikos Gorge to the mountain town of Kalavrita. The gorge is formed by fluvial incision through one of the uplifted fan deltas in the footwall of the now-active eastern Eliki fault. Ironically, the deposits of the Plio-Pleistocene fan delta were laid down by the same river that now incises and recycles sediments to produce the modern subsiding fan. The journey takes approximately 1.5 hours to Kalavrita or there is an intermediate stop with access to a local monastery. You can return immediately on the next train or spend some time in Kalvrita. During the train ride, you will see impressive views of the uplifted Gilbert fan-delta deposits in the cliffs of the heavily incised river gorge. There are 4–5 scheduled trains each day.

Stop 7.5 Eastern Eliki fault exposure and trench locations   Follow the Old National Road to 17.1 km beyond the NNR junction in Paralia Akrata. This is a few kilometres beyond the town of Diakofto and after a road bridge across the New National Road where the ONR sits to the south of the NNR. It is also 5.9 km from the turning to Trapeza (stop 7.3). The trench sites of Koukouvelas et al. (2001) are to the left of the ONR. Driving to 17.9 km brings you to the location of the McNeill et al. (2005b) trenches. All trenches described in the above publications are now filled in, at time of going to press. Additional trenches may have been excavated and may still be open.

Attempt to park on the left-hand side of the road to avoid crossing the road on foot. There is space for several cars and a single small coach to park along this stretch of road. At this particular location, one of the trenches exposed by Koukouvelas et al. (2001) is located but now filled in. This is a short walk from the edge of the road. Other trenches have been excavated east of this location by Koukouvelas et al. and by McNeill et al. (2005b) to the west, but most are now filled in. Be careful around and crossing this road – traffic is extremely fast through this section. Do not enter trenches (if open), as walls may be unstable. (Coordinates: 38°11'20"N, 22°9'55"E)

The eastern Eliki fault trace generally sits close to the base of slope, representing the transition from the footwall to hanging wall. The fault plane dips north at 50–60°, as measured on exposures of the bedrock (cemented deltaic conglomerate) fault plane at this location. Koukouvelas et al. (2001) suggest that some of the small (0.5 m) scarps in unconsolidated sediments observed at this location may be preserved from the 1861 earthquake.

However, McNeill et al. (2005b) suggest that many of these scarps may be a result of human excavation, observing no displacement beneath an example scarp during excavations. Uplifted marine terraces at this location (see Fig. 7.2) suggest that the fault footwall is uplifting at 1 mm a year (McNeill & Collier 2004).

Two groups have separately dug palaeoseismological trenches across traces of the Eliki fault. At 17.1 km, an open area south of the Old National Road and north of the base of slope (probably marking the fault itself) was chosen for several trenches by the University of Patras group, which excavate the hanging wall of the fault (Koukouvelas et al. 2001). These trenches revealed multiple fault splays at the surface. Koukouvelas et al. identify three displacement events in the past 2000 years (displacement per event 0.5–1.5 m). The youngest of these events may be the 1861 earthquake.

McNeill et al. (2005b) excavated trenches west of the Patras sites, at 17.9 km along the ONR. The sites are located close to a blue shrine on the north side of the road. Trench EET1 to the south of the road crossed a small scarp on an alluvial fan. However, no displaced sediments were found in this trench, supporting the view that many scarps in the area are not tectonic in origin. A second trench (EET1A) was excavated north of the road. Coarse-grain fluvial conglomerates are dominant and are thought to be deposits from the palaeo Kerinitis River when it flowed to the east along the range front. Several fault splays and also evidence for liquefaction were found in this trench (Fig. 7.10). The trench provides evidence for three earthquakes in the past 1500 years, probably including the 1861 earthquake, with timing constrained by radiocarbon dates. Recurrence intervals between earthquakes are 200–600 years. It is possible that not all earthquakes on the eastern Eliki fault are recorded in these trenches, because of multiple fault splays at the surface. A liquefied sand dyke at the north end of the trench is composed of structureless well sorted sand (Fig. 7.11). This dyke would have brought fluidized sand to the surface, possibly forming a sand volcano as observed following the 1861 earthquake by Schmidt (1879; Fig. 7.11). Dating of layers close to the dyke suggests that it formed during the 1861 event.

Stop 7.6 Kerinitis Gilbert fan delta, Kerinitis Gorge   Follow the ONR and take a left turn at 18.6 km from the NNR junction in Paralia Akrata, or 7.4 km beyond the Trapeza turnoff. This turning is just after the large bridge across the river and is signposted to Kato Pteri. Follow this road for 2.9 km and follow the road around to the left. Stop at 3.5 km. Plentiful parking for cars or coaches is available outside the tavernas. Only cars can continue around the traverse, although the many sharp bends make parking somewhat difficult in places.

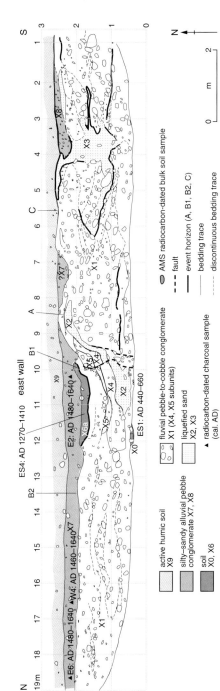

Figure 7.10 Trench log of palaeoseismological trench across the eastern Eliki fault, revealing displaced sedimentary layers (ages shown from radiocarbon dates) and liquefaction features (after McNeill et al. 2005b).

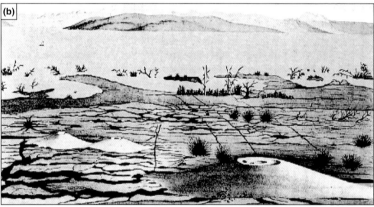

Figure 7.11 Liquefaction during Gulf of Corinth earthquakes. **(a)** A liquefied sand dyke identified in trench across eastern Eliki fault (Fig. 7.10) by McNeill et al. (2005b). **(b)** An illustration of sand volcanoes erupted during the 1861 earthquake sourced from a similar dyke (Schmidt 1879).

The stop is arranged as a hike covering about 3 km with moderate gradients. For the most spectacular overviews, take the metalled road from the tavernas. This soon bears left at a junction with a gravel track (the latter passable by car and providing the return route) to cross the Kerinitis River. Carry on ascending to the village, where a fountain provides welcome cool water. Turn right and carry on up hill. Once on the level, the road twists and turns parallel to the valley side. Fine views of the conglomerates making up the 6–700 m-high western gorge walls are soon available and, once the views open up to the southwest, various stops can be made to view the architecture of the deposit and its structural setting (Fig. 7.12). Carry on along the road, which eventually descends to cross the river once more and meets a gravel track. Walk back down stream to the coach along this track. Closer views of the cliff faces are available on the way, and under the main cliff it is possible to ascend to a couple of goat pens to see internal erosion surfaces and delta-front slump surfaces as described below. Hard hats should be worn on the gravel track below cliff sections. (Coordinates: 38°10′5″N, 22°7′6″E)

The brown-weathering conglomerates that make up the mountainside are of probable early Pleistocene age, although this is not directly dated but inferred by comparison with the age (from palynological evidence) of similar conglomerates in the next valley to the east, the Vouraikos River (Malartre et al. 2004). The thick (600 m) succession is faulted against Mesozoic limestone basement along the Mamussia–Pirgaki normal fault, whose prominent trace may be seen to the southwest of the panoramic view (Fig. 7.12). The fault is thought to be dead, active faulting having migrated north to the coastal Elike faults, as discussed previously on this leg of the itinerary. The detailed architecture revealed in the stupendous exposures consists of two broad elements. To the southwest are crudely stratified conglomerates thought to be fluviatile topset deposits. The unit is faulted on its southwest margin, with a small south back-tilt developed, more easily visible in some of the tributary gorges where true dips are revealed. It is assumed that this back-tilt was a response to sediment accommodation during active faulting. Closer examination along the gravel track reveals clast imbrication, cluster bedforms (clast logjams), crude cross stratification, erosion surfaces and shallow erosional scours. The clasts are subangular to sub-rounded, ranging from small pebbles to cobbles. The fluvial system responsible was probably a shallow braided channel network.

The architecture of the main cliff face is dominated by steeply dipping foreset deposits showing a complex interaction with the topsets. In dip sections, maximum foreset height is 600 m. When mapped out, the foresets show radial dips in the range 25–35° and, like the topsets, comprise mainly pebble- and cobble-grade conglomerates, although they are usually better sorted. Large syn-sedimentary slides bounded by curvilinear growth faults are also visible in dip sections through the foresets, whereas strike sections show the foreset units to be of lenticular geometry. The dominant mode of transport down the foresets appears to be that of grainflow avalanches and cohesionless debris flows, with both normal and reverse grading present in the individual foreset units. More massive beds may be attributable to hyperconcentrated flows. There is some evidence for

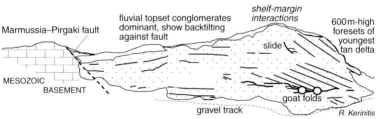

Figure 7.12 A panoramic view (looking NW) and simple sketch of the Kerinitis Gorge exposures through an early Pleistocene fan-delta system. The fan-delta deposits are faulted and backtilted against Mesozoic bedrock in the footwall to the inactive Mamussia–Pirgaki fault.

transformation of these grainflow units to turbidity-current deposits in exposures of bottomset beds at the far northern end of the whole exposure. Rotated slump blocks associated with erosion surfaces are clearly displayed in the exposures afforded by the goat folds under the main cliff face. These are accessible, with care and a prayer, from the gravel track on the return route.

Overall, the whole succession exposed here indicates aggradation of the fan-delta system. At the level of the main gravel track, a lower unit of foresets is overlain by coarse fluviatile topsets and then by lacustrine or floodplain marls at the level of the goat folds, establishing an overall progradational and aggradational cycle of fan delta out- and up-building. A period of relative base-level fall is recorded by a prominent erosion surface that incises into the topsets. Subsequent fan-delta progradation is recorded by a 200 m-thick forest wedge that advances to the northeast. It

is this unit that has distinctive upslope dipping wedges of conglomerates interpreted as slumped blocks. In detail, the many internal erosion surfaces that cut individual foreset and topset units in the central to northern parts of the exposure probably record erosion during periods of relative sea-level fall, with subsequent rising base level and highstand aggradation and progradation.

Stop 7.7 Agios Georgios: overview of Eliki and Aigion faults   At 21.8 km from the NNR junction, take a left turn off the ONR where signposted to Eliki, with a white shrine on your left. Follow this road for 0.6 km into the village square of Eliki and take a left turn. Follow this road for a further 2.7 km under the NNR bridge and along hairpins up the hillside. Just before the village of Keryneia, take a turning to the left onto a gravel road. Follow this for 0.9 km to a small chapel. Follow signs marked "ancient settlement" along this track. This site is suitable for cars and small groups only, because of a dirt road, steep climb and limited parking.

Limited parking is available for a few cars adjacent to the small chapel at the end of the gravel road. The viewpoint across the gulf and of Aigion from the Aigion fault is located at the chapel. Alternatively, walk east along the gravel road for another viewpoint towards the east along the eastern Eliki fault. The ancient acropolis site (high ground southeast of chapel) can also be explored. Be careful to avoid the cliff edges. (Coordinates: 38°12′29″N, 22°7′35″E)

This site sits on top of terraces within the uplifting footwall block of the western Eliki fault and is thought to be the location of the acropolis of the ancient city of Helike (Soter & Katsonopoulou 1998b). The main city was located on the coastal plain below (Soter & Katsonopoulou 1998a) and was mostly destroyed during the 373 BC earthquake. The foundations of a small Hellenistic temple and remains of a bronze statue have been excavated here. The site also has the correct morphology for a Mycenean acropolis. The materials exposed (Mycenean to Roman) suggest a substantial inhabitation for some period of time. At this position, we can also see the geomorphological effects of both the Eliki and Aigion fault systems and how they interact.

Terraces within the footwall of the western Eliki fault have been correlated with sea-level highstands and suggest an uplift rate of 1.25 mm a year (DeMartini et al. 2004). Farther west, uplift may be partially transferred to the Aigion fault to the north.

From the small chapel, a view to the north shows the Aigion fault with the city of Aigion sitting on top of the fault scarp. The Aigion fault extends east to the offshore continental shelf, where it displaces the sea floor (see Fig. 7.1; McNeill et al. 2005a). To the west, the fault extends to the Meganites River and farther west still is composed of several complex shorter segments (Palyvos et al. in press). The fault clearly decreases in displacement from west to east at the city. This displacement profile of maximum in the fault centre decreasing towards the fault tips, although expected for normal fault systems, is uncommon in established faults of the gulf. This evidence, coupled with the fault's position to the north and parallel to the Eliki fault, the limited topography of the fault footwall and the apparent cessation of activity on part of the western Eliki fault, all suggest that the Aigion fault is young and still growing. The fault may represent the process of fault activity stepping to the north, as has occurred to other fault systems in the past.

From the chapel, the Selinous River can be seen crossing the fan from west to east and entering the gulf. The river formerly entered the gulf to the north at the apex of the fan indicated by a bridge of Roman age. The river has gradually shifted south with time. This is thought to be attributable to the gradual tilting caused by uplift and growth of the Aigion fault and warping in the region between the Aigion and western Eliki faults.

The likely position of the now-buried city of Helike (following its destruction in 373 BC) can be seen below on the coastal plain to the south of the Selinous River (Soter & Katsonopoulou 1998a).

Walking farther along the track behind the chapel towards the east; additional views of the Eliki fault system (Fig. 7.13) and rest of the gulf can be seen. The archaeological site can be investigated if time permits, but few remains are visible.

Figure 7.13 View east along eastern Eliki fault from Aigios Konstantinos. The clear fault boundary between the uplifting mountainous footwall block to the south and the subsiding coastal plain of the hanging-wall block to the north can be seen.

Stop 7.8 Western Eliki fault limestone scarp For cars, at 22.4 km from the NNR junction and 0.6 km beyond the Eliki turnoff (stop 7.7) take the turning to the left signposted to Kerynia. At 0.3 km along this road, turn right; at 1.3 km turn left under bridge. Just under the bridge take the lower elevation of two roads to the left. Travel east along this gravel road for about 0.5 km. Park at this location. Coaches should travel 24.1 km along the ONR from NNR junction in Paralia Akrata. A large and complex junction is located here with a wood yard on the left and a garden centre on the right, allowing access onto the road into Aigion. Take the road on the other side (left or south) of the ramp, which leads up to the Aigion road and the continuation of the ONR. This involves turning left and stopping, then crossing oncoming traffic to the other (south) side. Travel 1.5 km along this road to a bridge under the New National Road (you will travel south and back towards the east along this road). Park a coach at the side of this road (allowing other vehicles to pass).

Parking for cars is available along the gravel track. A coach party should park on the road before the bridge under the NNR and walk from here. Walk east along the gravel track about 0.5 km to outcrop of bedrock (limestone) fault scarp of the western Eliki fault exposed by a landslide on your right or to the south. You are located between the fault scarp and the NNR. From the coach, walk under the bridge for the NNR, turn left onto the lower-elevation gravel road and walk for about 0.5 km. Do not venture onto the New National Road. (Coordinates: 38°12′50″N, 22°7′3″E)

141

The limestone fault plane of the western Eliki fault is exposed at this location by a large landslide. This location is slightly elevated above the coastal plain and the actual active fault trace may be located slightly farther north.

The fault plane dips 55–60°N and displays many classic features. This is an excellent site for students to practise taking measurements with a compass. The fault zone is composed of brecciated limestone with breccia clasts between millimetres and centimetres in diameter. Slickenside striations are almost parallel to the dip of the fault plane indicating pure dip slip (i.e. no strike-slip component; Fig. 7.14). The fault plane is also strongly corrugated. Excellent examples of raised tool tracks or drag marks from breccia clasts show the direction of motion on the fault, confirming normal displacement (Fig. 7.15).

Turning away from the fault plane, the Aigion fault and city of Aigion can be seen from this location to the north and northwest beyond the modern village of Eliki. Details of activity of this fault, the western Eliki fault, and the interaction between the two fault systems are discussed at stop 7.3.

Stop 7.9 Aigion town, earthquake damage from 1995   Continue along the ONR into Aigion town approximately 27 km from the NNR junction in Paralia Akrata. Shortly after the Eliki and Keryneia turnoffs of stops 5.3 and 7.6, join the main road to Aigion at the junction described in stop 7.8 for coaches. This involves driving to the stop sign, turning right onto the apparent continuation of the road you have been driving on, and then bearing to the right to join the road into Aigion from the NNR exit. Once driving into Aigion (along ONR in the southeast corner of Fig. 7.16), follow the map provided to find locations where earthquake damage from 1995 can still be observed. Parking in the town is difficult; it is best to park away from the centre and walk to the sites described here using the map of Figure 7.16.

The city of Aigion suffered significant damage during the 15 July 1995 earthquake. Much of the damage is still evident today (at the time of going to press). Although the earthquake produced a few centimetres of displacement on the Aigion fault, the source of the earthquake is thought to be a fault off shore to the northeast. Damage appears to have been greatest close to the terrace edge marking the top of the Aigion fault scarp on the north side of the town. It is thought that this free face caused amplification of ground motions. Peak accelerations reached 0.5 g in parts of the city (Koukouvelas 1998).

Many of the damaged buildings described here are located around a church (Fig. 7.16) near the terrace edge, which is now supported by scaffolding and buttresses (Fig. 7.17). Building damage includes roofs, cracks

Figure 7.14 Bedrock fault scarp of the western Eliki fault showing pure dip-slip striations on the fault plane. The fault plane dips at 55–60°.

to masonry around window and door frames, damage to balconies and roof fittings, and total failure of walls (Fig. 7.18). Gaps between buildings indicate where collapse occurred or buildings were beyond repair and pulled down. Some of the buildings are now being renovated, but many have been abandoned since the earthquake. The building damage suggests intensities of VI–VIII on the modified Mercalli scale.

Palaeoseismological trenching of the Aigion fault west of the city of Aigion (Pantosti et al. 2004) revealed three displacement events in the thousand years prior to 1888 (therefore not including the small displacement of the 1995 earthquake), with average recurrence intervals of about 360 years. These intervals are similar to other faults within the gulf.

Figure 7.15 Drag marks, striations and sheared fault breccia clasts on the western Eliki fault scarp.

Further reading

The most comprehensive recent accounts of the tectonics and structure of the western Gulf of Corinth appear in the 2004 special volume of *Comptes Rendus Geoscience* (**336**, 235–485) devoted to results of the Franco-Greek Corinth Rift Project (summarized by Cornet et al. 2004). More focused papers outlining recent work on onshore and offshore geology are by

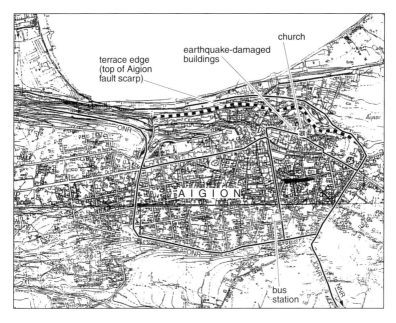

Figure 7.16 Aigion city, highlighting areas where buildings with damage from the 1995 earthquake are still preserved (at time of going to press).

Figure 7.17 Church under reconstruction following 1995 earthquake damage.

Figure 7.18 Buildings showing typical earthquake damage.

Stefatos et al. (2002) and McNeill et al. (2005a), and the spectacular uplifted Gilbert fan-delta deposits are well documented by Ori (1989), Dart et al. (1994) and Malartre et al. (2004).

Glossary

alluvial deposits, alluvium, fluvial Detrital material transported by a river channel and deposited at points along the flood plain.

colluvial Hillslope deposits, the products of hillslope erosion and transport, e.g. by soil creep, rain splash and rain wash, also landslides and mass wastage.

colluvium Generally unsorted and angular sediment, derived from hillslopes and formed by creep and local surficial runoff.

core complex Mid-to-lower crust exposed by exhumation caused by the extension of thickened hot lithosphere along low angle normal faults (detachments).

cross stratification Inclined strata deposited by grain avalanching causeed by unidirectional, combined or oscillatory current motions. It can range from centimetre scale to hundreds of metres.

debris flow A cohesive mass flow that can support a mass of anomalously large clasts or blocks that "float" within the flow.

facies The overall characteristics of a sediment. It may be used either in a descriptive sense (e.g. sandstone facies) or in an interpretive sense (e.g. shallow marine facies).

fan delta Characteristic base-of-slope (often fault-bounded) fan-shape sediment accumulation that enters a lake or the sea.

footwall The upthrown side of a normal fault whose net upwards motion during an earthquake is termed footwall uplift.

Gilbert-type fan delta A subset of the fan-delta family, with a characteristic threefold division to its structure, with subhorizontal, river-deposited topsets passing into steeply inclined ($>20°$),

avalanche grainflow foresets that asymptote down slope into lower angle and typically finer-grain bottom sets.

graben A sedimentary basin bounded by two, opposed normal faults which dip towards the basin.

half-graben A sedimentary basin, asymmetric in cross section, the floor of which tilts down towards a single normal fault.

hanging wall The downthrown side of a normal fault whose net downward motion during an earthquake is termed hanging-wall subsidence.

hazard Used strictly, it is the statistical chance of having an earthquake of a particular magnitude at a particular location.

horst An upthrown block between two, opposed normal faults which dip away from the block.

imbrication A sedimentary fabric in which particles lie such that the plane including their long axis is tilted up stream in a stable hydrodynamic arrangement.

intensity (modified Mercalli intensity scale) Scaled I–XII, is a measure of the physical effects of an earthquake at a particular location, can be contoured. The following is much abridged:

  I Not felt. Marginal and long-period effects of large earthquakes.

  II Felt by persons at rest, on upper floors, or favourably placed.

  III Felt indoors. Hanging objects swing. Vibration like passing of light trucks.

  IV Hanging objects swing. Vibration like passing of heavy trucks. Standing cars rock. Windows, dishes, doors rattle.

  V Felt outdoors. Sleepers wakened. Liquids disturbed. Doors swing, close, open.

  VI Felt by all. Many frightened and

run outdoors. Persons walk unsteadily. Windows, dishes, glassware broken. Pictures off walls. Furniture moved or overturned.

VII Difficult to stand. Noticed by drivers. Hanging objects quiver. Furniture broken. Damage to masonry. Waves on ponds, water turbid with mud.

VIII Steering of cars affected. Damage to masonry, partial collapse. Twisting, fall of chimneys, factory stacks, monuments, towers, elevated tanks. Cracks in wet ground and on steep slopes.

IX General panic. Masonry heavily damaged, sometimes with complete collapse. General damage to foundations. Serious damage to reservoirs. Conspicuous cracks in ground.

X Most masonry and frame structures destroyed with their foundations. Some well built wooden structures and bridges destroyed. Large landslides.

XI Rails bent greatly. Underground pipelines completely out of service.

XII Damage nearly total. Large rock masses displaced. Lines of sight and level distorted. Objects thrown into the air.

listric Any normal fault whose dip decreases smoothly with depth.

lithospheric extension The process by which lithosphere is thinned, the upper crust mostly by faulting (brittle failure) and the rest by ductile (plastic) flow.

magnitude An instrument based measurement of the size of an earthquake (i.e. does not consider effects). Basically it is $\log_{10}$ (max. amplitude in microns of seismic wave on "standard seismograph" at 100 km from epicentre). Magnitude 8 is enormous (*c.* two events a year, worldwide), 6 would certainly be severe and could cause lots of damage (*c.* a hundred a year), 4 could be distinctly felt (15 000 a year).

mass flow Concentrated, sediment-rich flows in which turbulence is inhibited.

neotectonic Literally, the "new tec-tonics", i.e. the current tectonic regime.

normal fault Any fault that throws down or obliquely down in the direction of fault dip; associated with stretching of the Earth's crust.

post-rift The part of a basin's history that occurred after active faulting had ceased (although there might still have been a topographic expression of older faults and subsidence might still have been occurring, due to cooling of the Earth's lithosphere).

risk = hazard x vulnerability.

rollover The convex-upwards form assumed by strata as they dip progressively more steeply into the plane of a normal fault, particularly those faults with listric geometry.

sequence An interval of genetically related strata deposited under conditions of highstand, rising or falling sea level.

sequence boundary A surface, often an erosion surface or disconformity, marking a change in environmental conditions such as a lowstand in sea level.

sequence stratigraphy The methodology by which a basin fill is separated into a series of genetically related sequences and bounding surfaces for purposes of correlation. This approach allows stratigraphical predictions to be made, on the assumption that all discontinuities and unconformities relate to changes in sea level.

slab/trench rollback The oceanward collapse of a subducting oceanic lithospheric slab that causes back-arc extension in the overriding plate.

streamflow Aqueous flows with dilute sediment loads.

syn-rift The part of a basin's history that occurred when rift faulting was active.

tilt block Volume of crust bounded by active normal faults dipping in the same direction.

transfer zone The gap between two offset normal-fault segments.

turbidite Deposit of a turbidity current.

turbidity current or flow High energy, turbulent flows, often forming density underflows in lake or marine settings when the combined density of freshwater + sediment coming out of a river exceeds the density of the basin's water column.

vulnerability Takes into account design of buildings, foundations, local geology etc. to provide the relative potential for damage cause by an earthquake.

Bibliography

Abercrombie, R. E., I. C. Main, A. Douglas, P. W. Burton 1995. The nucleation and rupture processes of the 1981 Gulf of Corinth earthquakes from deconvolved broad-band data. *Geophysical Journal International* **120**, 393–405.

Adey, W. H. 1986. Coralline algae as indicators of sea level. In *Sea-level research: a manual for the collection and evaluation of data*, O. van de Plassche (ed.), 229–80. Norwich: GeoBooks.

Ambraseys, N. N. & J. A. Jackson 1997. Seismicity and strain in the Gulf of Corinth (Greece) since 1694. *Journal of Earthquake Engineering* **1**, 433–74.

Andrews, J. E., C. Portman, P. J. Rowe, M. R. Leeder, J. D. Kramers 2007. Suborbital sea-level change in early MIS 5e: new evidence from the Gulf of Corinth, Greece. *Earth and Planetary Science Letters* **259**, 457–68.

Armijo, R., B. Meyer, G. C. P. King, A. Rigo, D. Papanastassiou 1996. Quaternary evolution of the Corinth Rift and its implications for the late Cenozoic evolution of the Aegean. *Geophysical Journal International* **126**, 11–53.

Avallone, A. and 9 other authors 2004. Analysis of eleven years of deformation measured by GPS in the Corinth Rift Laboratory area. *Comptes Rendus Geoscience* **336**, 301–311.

Bentham, P. R. E., R. E. Ll. Collier, R. L. Gawthorpe, M. R. Leeder, S. Prossor, C. P. Stark 1991. Tectono-sedimentary development of an extensional sedimentary basin: the Neogene Megara Basin, Greece. *Geological Society of London, Journal* **148**, 923–34.

Bernard, P. Briole, B. Meyer and 5 other authors 1997. The Ms = 6.2 June 15 1995 Aigion earthquake (Greece): evidence for low-angle normal faulting in the Corinth rift. *Journal of Seismology* **1**, 131–50.

Bornovas, J., P. Gaitanakis, A. Spiridopoulos 1984. *Geological map of Greece, 1:50 000, Perachora sheet*. Athens: IGME.

Brooks, M. & G. Ferentinos 1984. Tectonics and sedimentation in the Gulf of Corinth and the Zakynthos and Kefallinia channels, western Greece. *Tectonophysics* **101**, 25–54.

Briole, P., A. Rigo, H. Lyon-Caen and 7 other authors 2000. Active deformation of the Corinth rift, Greece: results from repeated Global Positioning System surveys between 1990 and 1995. *Journal of Geophysical Research* **105**, 25605–25625.

Chappell, J., A. Omura, T. Esat and 4 other authors 1996. Reconciliation of late Quaternary sea levels derived from coral terraces at Huon Peninsula with deep-sea oxygen isotope records. *Earth and Planetary Science Letters* **141**, 227–36.

Chappell, J. & N. J. Shackleton 1986. Oxygen isotopes and sea level. *Nature* **324**, 137–40.

Clarke, P. J., R. R. Davies, P. C. England and 8 other authors 1997. Geodetic estimate of seismic hazard in the Gulf of Korinthos. *Geophysical Research Letters* **24**, 1303–306.

Collier, R. E. Ll. 1990. Eustatic and tectonic controls upon Quaternary coastal sedimentation in the Corinth Basin. *Geological Society of London,*

Journal **147**, 301–314.

Collier, R. E. Ll & C. J. Dart 1991 Neogene to Quaternary rifting, sedimentation and uplift in the Corinth Basin, Greece. *Geological Society of London, Journal* **148**, 1049–1065.

Collier, R. E. Ll & J. Thompson 1991. Transverse and linear dunes in an Upper Pleistocene marine sequence, Corinth Basin, Greece. *Sedimentology* **38**, 1021–1040.

Collier, R. E. Ll., M. R. Leeder, P. J. Rowe, T. C. Atkinson 1992. Rates of tectonic uplift in the Corinth and Megara basins, central Greece. *Tectonics* **11**, 1159–67.

Collier, R. E. Ll., D. Pantosti, G. Daddezio, P. M. De Martini, E. Masana, D. Sakellariou 1998. Paleoseismicity of the 1981 Corinth earthquake fault: seismic contribution to extensional strain in central Greece and implications for seismic hazard. *Journal of Geophysical Research* **103**, 30001–30019.

Collier, R. E. Ll., M. R. Leeder, M. Trout, G. Ferentinos, E. Lyberis, G. Papatheodorou 2000. High sediment yields and cool, wet winters during the last glacial lowstand, northern Mediterranean. *Geology* **28**, 999–1002.

Cornet, F. H., P. Bernard, I. Moretti 2003. The Corinth Rift Laboratory. *Comptes Rendus Geoscience* **336**, 235–41.

Cowie, P. A. & G. P. Roberts 2001. Constraining slip rates and spacings for active normal faults. *Journal of Structural Geology* **23**, 1901–915.

Dart, C. J., R. E. Ll. Collier, R. L. Gawthorpe, J. V. A. Keller, G. Nichols 1994. Sequence stratigraphy of (?) Pliocene–Quaternary synrift, Gilbert-type fan deltas, northern Peloponnisos, Greece. *Marine and Petroleum Geology* **11**, 545–60.

Davies, R., P. England, B. Parsons, H. Billiris, D. Paradissis, G. Veis 1997.

Geodetic strain of Greece in the interval 1892–1992. *Journal of Geophysical Research* **102**, 24571–24588.

Demartini, P. M., D. Pantosti, N. Palyvos, F. Lemeille, L. McNeill, R. Collier 2004. Slip rates of the Aigion and Eliki faults from uplifted marine terraces, Corinth Gulf, Greece. *Comptes Rendus Geoscience* **336**, 325–34.

Dia, A. N., A. S. Cohen, R. K. O'Nions, J. A. Jackson 1997. Rates of uplift investigated through ^{230}Th dating in the Gulf of Corinth (Greece). *Chemical Geology* **138**, 171–84.

Dixon, J. E. & A. H. F. Robertson (eds) 1984. *The geological evolution of the eastern Mediterranean*. Special Publication 17, Geological Society, London.

Duermeijer, C. E., W. Krijgsman, C. G. Langereis, J. E. Meulenkamp, M. V. Triantaphyllou, W. J. Zachariasse 1999. A Late Pleistocene clockwise rotation phase of Zakynthos (Greece) and implications for the evolution of the western Aegean arc. *Earth and Planetary Science Letters* **173**, 315–31.

Duermeijer, C. E., M. Nyst, P. Th. Meijer, C. G. Langereis, W. Spakman 2000. Neogene evolution of the Aegean arc: palaeomagnetic and geodetic evidence for a rapid and young rotation phase. *Earth and Planetary Science Letters* **176**, 509–525.

Emeis, K-C., H. Schulz, U. Struck, M. Rossignol-Strick, H. Erlenkeuser, M. W. Howell and 6 other authors 2003. Eastern Mediterranean surface water temperatures and σ^{18}O composition during deposition of sapropels in the late Quaternary. *Paleoceanography* **18**(1), 1005, DOI:10.1029/2000PA000617, 5-1-5-18.

Faccenna, C., T. W. Becker, F. P. Lucente, L. Jolivet, F. Rossetti 2001. History of subduction and back-arc extension in the central

Mediterranean. *Geophysical Journal International* **145**, 809–820.

Ferentinos, G., G. Papatheodorou, M. B. Collins 1988. Sediment transport processes on an active submarine fault escarpment: Gulf of Corinth, Greece. *Marine Geology* **83**, 43–61.

Freyberg, B. von 1973. Geologie des Isthmus von Korinth. *Erlanger Geologische Abhandlungen* **95**. Junge und Sohn, Universitäts-Buchdruckerei, Erlangen.

Frogley, M. R., P. C. Tzedakis, T. H. E. Heaton 1999. Climate variability in NW Greece during the last interglacial. *Science* **285**, 1886–9.

Gawthorpe, R. L., A. J. Fraser, R. E. Ll. Collier 1994. Sequence stratigraphy in active extensional basins: implications for the interpretation of ancient basin fills. *Marine and Petroleum Geology* **11**, 642–58.

Goldsworthy, M. & J. Jackson 2001. Migration of activity within normal fault systems: examples from the Quaternary of mainland Greece. *Journal of Structural Geology* **23**, 489–506.

Graves, R. 1992. *The Greek myths*. Harmondsworth: Penguin.

Hatzfeld, D., V. Karakostas, M. Ziaza, I. Kassaras, E. Papadimitriou, K. Makropoulos, N. Voulgaris, C. Papaioannou 2000. Microseismicity and faulting geometry in the Gulf of Corinth (Greece). *Geophysical Journal International* **141**, 438–56.

Houghton, S. L., G. P. Roberts, D. Papanikolaou, J. M. McArthur, M. A. Gilmour 2003. New 234U–230Th coral dates from the western Gulf of Corinth: implications for extensional tectonics. *Geophysical Research Letters* **30**, 19, 2013, DOI: 10.1029/ 2003GL018112.

Hubert, A., G. C. P. King, R. Armijo, B. Meyer 1996. Fault reactivation, stress interaction and rupture propagation in the 1981 Corinth earthquake sequence. *Earth and Planetary Science Letters* **142**, 573–86.

Imbrie, J., J. D. Hays, D. G. Martinson and 6 other authors 1984. The orbital theory of Pleistocene climate: support from a revised chronology of the marine [18]O records. In *Milankovitch and climate, part 1*, A. L. Berger et al. (eds), 269–305. Dordrecht: Reidel.

Jackson, J. A. & D. P. McKenzie 1983. The geometrical evolution of normal fault systems. *Journal of Structural Geology* **11**, 15–36.

Jackson J. A. & D. P. McKenzie 1988. The relationship between plate motions and seismic moment tensors, and the rates of active deformation in the Mediterranean and Middle East. *Geophysical Journal* **93**, 45–73.

Jackson, J. A. & N. J. White 1989. Normal faulting in the upper crust: observations from regions of active extension. *Journal of Structural Geology* **11**, 15–36.

Jackson, J. A., J. Gagnepain, G. Houseman, G. C. P. King, P. Papadimitriou, C. Soufleris, J. Virieux 1982a. Seismicity, normal faulting, and the geomorphological development of the Gulf of Corinth (Greece): the Corinth earthquakes of February and March 1981. *Earth and Planetary Science Letters* **57**, 377–97.

Jackson, J. A., G. C. P. King, C. Vita-Finzi 1982b. The neotectonics of the Aegean: an alternative view. *Earth and Planetary Science Letters* **61**, 303–318.

Keraudren, B. & D. Sorel 1987. The terraces of Corinth (Greece) – a detailed record of eustatic sea-level variations during the last 500 000 years. *Marine Geology* **77**, 99–107.

Kershaw, S. & L. Guo 2001. Marine

notches in coastal cliffs: indicators of relative sea-level change, Perachora Peninsula, central Greece. *Marine Geology* **179**, 213–28.

Kershaw, S. & L. Guo 2003. Pleistocene cyanobacterial mounds in the Perachora Peninsula, Gulf of Corinth, Greece: structure and applications to interpreting sea-level history and terrace sequences in an unstable tectonic setting. *Palaeogeography, Palaeoclimatology, Palaeoecology* **193**, 503–514.

King, G. C. P., Z. X. Ouyang, P. Papadimitriou and 6 other authors 1985. The evolution of the Gulf of Corinth (Greece): an aftershock study of the 1981 earthquakes. *Royal Astronomical Society, Geophysical Journal* **80**, 677–93.

King, T. 1998. *Mechanisms of isostatic compensation in areas of lithospheric extension: examples from the Aegean.* PhD thesis, University of Leeds.

Koukouvelas, I. K. & T. Doutsos 1996. Implications of structural segmentation during earthquakes: the 1995 Egion earthquake, Gulf of Corinth, Greece. *Journal of Structural Geology* **18**, 1381–8.

Koukouvelas, I. K. 1998. The Egion fault, earthquake-related and long-term deformation, Gulf of Corinth, Greece. *Journal of Geodynamics* **26**, 501–513.

Koukouvelas, I. K., L. Stamatopoulos, D. Katsonopoulou, S. Pavlides 2001. A palaeoseismological and geoarchaeological investigation of the Eliki fault, Gulf of Corinth, Greece. *Journal of Structural Geology* **23**, 531–43.

Kumerics, C., U. Ring, S. Brichau, J. Glodny, P. Monie 2005. The extensional Messaria shear zone and associated brittle detachment faults, Aegean Sea, Greece. *Geological Society of London, Journal* **162**, 701–722.

Laj, C., M. Jamet, D. Sorel, J-P. Valente 1982. First palaeomagnetic results from Mio-Pliocene series of the Hellenic sedimentary arc. *Tectonophysics* **86**, 45–67.

Leeder, M. R., M. Seger, C. P. Stark 1991. Sedimentology and tectonic geomorphology adjacent to active and inactive normal faults in the Megara Basin and Alkyonides Gulf, central Greece. *Geological Society of London, Journal* **148**, 331–43.

Leeder, M. R., T. Harris, M. J. Kirkby 1998. Sediment supply and climate change: implications for basin stratigraphy. *Basin Research* **10**, 7–18.

Leeder, M. R., R. E. Ll. Collier, L. H. A. Aziz and 4 other authors 2002. Tectono-sedimentary processes along an active marine/lacustrine half-graben margin: Alkyonides Gulf, E. Gulf of Corinth, Greece. *Basin Research* **14**, 25–41.

Leeder, M. R., L. C. McNeill, R. E. Ll. Collier and 4 other authors 2003. Corinth rift margin uplift: new evidence from Late Quaternary marine shorelines. *Geophysical Research Letters* **30**, 1611–14.

Leeder, M. R., C. Portman, J. E. Andrews and 6 other authors 2005. Normal faulting and crustal deformation, Alkyonides Gulf and Perachora Peninsula, eastern Gulf of Corinth rift, Greece. *Geological Society of London, Journal* **162**, 549–61.

Le Pichon X. & J. Angelier 1979. The Hellenic arc and trench system: a key to the neotectonic evolution of the eastern Mediterranean area. *Tectonophysics* **60**, 1–42.

Lister, G. S. 1984. Metamorphic core complexes or Cordilleran type in the Cyclades, Aegean Sea, Greece. *Geology* **12**, 221–5.

Malartre, F., M. Ford, E. A. Williams 2004. Preliminary biostratigraphy and 3-D geometry of the Vouraikos Gilbert-type fan delta, Gulf of Corinth, Greece. *Comptes Rendus Geoscience* **336**, 269–80.

Marinatos, S. 1960. Helice: a submerged town of Classical Greece. *Archaeology* **13**, 186–93.

McCluskey, S., S. Balassanian, J. Barka and 25 other authors 2000. Global positioning system constraints on plate kinematics and dynamics in the eastern Mediterranean and Caucusus. *Journal of Geophysical Research* **105**, 5695–719.

McKenzie, D. P. 1972. Active tectonics of the Mediterranean region. *Royal Astronomical Society, Geophysical Journal* **30**, 109–185.

—— 1978. Some remarks on the development of sedimentary basins. *Earth and Planetary Science Letters* **40**, 25–32.

McKenzie, D. P. & J. A. Jackson 1986. A block model of distributed deformation by faulting. *Geological Society of London, Journal* **143**, 249–53.

McMurray, L. S. & R. L. Gawthorpe 2000. Along-strike variability of forced regressive deposits: Late Quaternary, northern Peloponnisos, Greece. In *Sedimentary responses to forced regressions*, D. Hunt & R. L. Gawthorpe (eds), 363–77. Special Publication 172, Geological Society, London.

McNeill, L. C. & R. E. Ll. Collier 2004. Footwall uplift rates of the eastern Eliki fault, Gulf of Corinth, Greece, inferred from Holocene and Pleistocene terraces. *Geological Society of London, Journal* **161**, 81–92.

McNeill, L., C. Cotterill, A. Stefatos and 6 other authors 2005a. Active faulting within the offshore western Gulf of Corinth, Greece: implications for models of continental rift deformation. *Geology* **33**, 241–4.

McNeill, L., R. E. L. Collier, D. Pantosti, P. M. Demartini, G. D'Adezio 2005b. Recent history of the eastern Eliki fault, Gulf of Corinth: geomorphology, paleoseismology and impact on paleoenvironments. *Geophysical Journal International* **161**, 154–66.

Moretti, I., D. Sakellariou, V. Lykousis, L. Micarelli 2003. The Gulf of Corinth: an active half graben? *Journal of Geodynamics* **36**, 323–40.

Morewood, N. C. & G. P. Roberts 1999. Lateral propagation of the surface trace of the South Alkyonides normal fault segment, central Greece: its impact on models of fault growth and displacement-length relationships. *Journal of Structural Geology* **21**, 635–52.

—— 2002. Surface observations of active normal fault propagation: implications for growth. *Geological Society of London, Journal* **159**, 263–72.

Mouyaris, N., D. Papastamatiou, C. Vita-Finzi 1992. The Helice fault? *Terra Nova* **4**, 124–9.

Noller, J., L. Wells, E. Reinhart, R. Rothaus 1997. Subsidence of the harbour of Kenchreai, Saronic Gulf, Greece, during the earthquakes of AD 400 and AD 1928. *EOS (Fall Meeting abstract volume), Transactions of the American Geophysical Union* **78**, F636.

Ori, G. G. 1989. Geological history of the extensional basin of the Gulf of Corinth (?Miocene–Pleistocene), Greece. *Geology* **17**, 918–21.

Palyvos, N., D. Pantosti, P. M. De Martini, F. Lemeille, D. Sorel, K. Pavlopoulos 2005. The Aigion–Neos Erineos coastal normal fault system (Western Corinth Gulf rift, Greece): geomorphological signature, recent earthquake history and evolution. *Journal of Geophysical Research B* **110**(9), 1–15.

Pantosti, D., P. M. Demartini, I. Koukouvelas and 5 other authors 2004. Paleoseismological investigations of the Aigion fault (Gulf of Corinth, Greece). *Comptes Rendus Geoscience* **336**, 335–42.

Papatheodorou, G. & G. Ferentinos 1993. Sedimentation processes and basin-filling depositional architecture in an active asymmetric graben: Strava graben, Gulf of Corinth, Greece. *Basin Research* **5**, 235–53.

Papatheoderou, G. & G. Ferentinos 1997. Submarine and coastal sediment failure triggered by the 1995, Ms = 6.1 Aigion earthquake, Gulf of Corinth, Greece. *Marine Geology* **137**, 287–304.

Papazachos, B. C. & C. B. Papazachos 1989. *The earthquakes of Greece* [in Greek]. Thessaloniki: Ziti.

Perissoratis, C., D. Mitropoulos, L. Angelopoulos 1984. The role of earthquakes in inducing sediment mass movements in the eastern Korinthiakos Gulf: an example from the February 24–March 4, 1981 activity. *Marine Geology* **55**, 35–45.

Perissoratis, C., D. Mitropoulos, L. Angelopoulos 1986. Marine geological research at the eastern Corinthiakos Gulf. In *Geological and Geophysical Research, Special Issue*, 381–401. Athens: IGME.

Perissoratis, C., D. J. W. Piper, V. Lykousis 2000. Alternating marine and lacustrine sedimentation during late Quaternary in the Gulf of Corinth rift basin, central Greece. *Marine Geology* **167**, 391–411.

Pirazzoli, P. A., S. C. Stiros, M. Arnold, J. Laborel, F. Laborel-Deguen, S. Papageorgiou 1994. Episodic uplift deduced from Holocene shorelines in the Perachora Peninsula, Corinth area, Greece. *Tectonophysics* **229**, 201–209.

Pirazzoli, P. A., S. C. Stiros, M. Fontugne, M. Arnold 2005. Holocene and Quaternary uplift in the central part of the southern coast of the Corinth Gulf (Greece). *Marine Geology* **212**, 35–44.

Portman, C., J. E. Andrews, P. J. Rowe, M. R. Leeder, J. Hoogewerff 2005. Submarine-spring controlled calcification and growth of large *Rivularia* bioherms, Late Pleistocene (MIS 5e), Gulf of Corinth, Greece. *Sedimentology* **52**, 441–65.

Powell, D. 1957. *An affair of the heart*.

London: Hodder & Stoughton.

Powell, D. 1967. The mirror of the present. *Classical Association, Proceedings* **64**, 1–12.

Richter, D. K., A. Herforth, E. Ott 1979. Pleistozane, brackische Blaugrunalgenriffe mit *Rivularia haematites* auf der Perachorahalbinsel bei Korinth (Greichenland). *Neues Jahrburgh fur Geologie und Palaontologie – Abhandlungen* **159**, 14–40.

Richter, D. K., H. Krampitz, R. Schillings 1989. Recent aragonite cement in a marine-meteoric mixing zone from the northern coast of the Perachora Peninsula near Corinth (Greece). *Zentralblatt Geologie und Paläontologie* **9/10**, 1369–82.

Riding, R. 1982. Cyanophyte calcification and changes in ocean chemistry. *Nature* **299**, 814–15.

Ring, U. & P. W. Layer 2003. High-pressure metamorphism in the Aegean, eastern Mediterranean: underplating and exhumation from the late Cretaceous until the Miocene to Recent above the retreating Hellenic subduction zone. *Tectonics* **22**, DOI: 10.1029/20010TC001350.

Roberts, G. P. 1996. Noncharacteristic normal faulting surface ruptures from the Gulf of Corinth, Greece. *Journal of Geophysical Research* **101**, 25255–67.

Roberts, G. P. & I. Koukouvelas 1996. Structural and seismological segmentation of the Gulf of Corinth fault system: implications for models of fault growth. *Annales Geofisica* **39**, 619–46.

Roberts, G. P. & I. S. Stewart 1994. Uplift, deformation and fluid involvement within an active normal fault zone in the Gulf of Corinth, Greece. *Geological Society of London, Journal* **151**, 531–42.

Robertson, A. H. F., P. H. Clift, P. J. Degnan, G. Jones 1991. Palaeogeographic and palaeotectonic evolution of the eastern

Mediterranean NeoTethys. *Palaeogeography, Palaeoclimatology, Palaeoecology* **87**, 289–343.

Rubie, S. & R. Van Der Hilst 2001. Processes and consequences of deep subduction: introduction. *Physics of the Earth and Planetary Interiors* **127**, 1–7.

Sachpazi, M., C. Clement, M. Laigle, A. Hirn, N. Roussos 2003. Rift structure, evolution and earthquakes in the Gulf of Corinth from reflection seismic images. *Earth and Planetary Science Letters* **216**, 243–57.

Sakellariou, D., V. Lykousis, D. Papanikolaou 1998. Neotectonic structure and evolution of the Gulf of Alkyonides, central Greece. *Geological Society of Greece, Bulletin* **32**, 241–50.

Schmidt, J. F. J. 1879. *Studien uber Erdbeben*. Leipzig: Carl Scholtze.

Soter, S. 1998. Holocene uplift and subsidence of the Helike Delta, Gulf of Corinth, Greece. In *Coastal tectonics*, I. S. Stewart & C. Vita-Finzi (eds), 41–56. Special Publication 146, Geological Society, London.

Soter, S. & D. Katsonopoulou 1998a. The search for ancient Helike 1988–1995: geological, sonar, and borehole studies. In *Ancient Helike and Aigialeia*, D. Katsonopulou, S. Soter, D. Schilardi (eds), 67–116. Proceedings of the Second International Conference, Helike Society, Athens.

Soter, S. & D. Katsonopoulou 1988b. Geophysical observations on the hill of Agios Georgios, possible site of the acropolis of Helike. In *Ancient Helike and Aigialeia*, D. Katsonopulou, S. Soter, D. Schilardi (eds), 117–130. Proceedings of the Second International Conference, Helike Society, Athens.

Spakman, W., M. Wortel, N. Vlaar 1988. The Hellenic subduction zone: a tomographic image and its geodynamic implications. *Geophysical Research Letters* **15**, 60–63.

Stefatos, A., G. Papatheodorou, G. Ferentinos, M. Leeder, R. Collier 2002. Seismic reflection imaging of active offshore faults in the Gulf of Corinth: their seismotectonic significance. *Basin Research* **14**, 487–502.

Stewart, I. S. 1996. Holocene uplift and palaeoseismicity on the Eliki fault, Western Gulf of Corinth, Greece. *Annali di Geofisica* **39**, 575–88.

Stewart, I. S. & C. Vita-Finzi 1996. Coastal uplift on active normal faults: the Eliki fault, Greece. *Geophysical Research Letters* **23**, 1853–6.

Taymaz, T., J. A. Jackson, D. P. McKenzie 1991. Active tectonics of the north and central Aegean. *Geophysical Journal International* **106**, 433–90.

Theodoropoulos, D. 1968. *Stratigraphie und Tectonik des Isthmus von Megara (Griechenland)*. Erlangen: Junge.

Thompson, W. G. & S. L. Goldstein 2005. Open-system coral ages reveal persistent suborbital sea-level cycles. *Science* **308**, 401–404.

Tiberi, C., H. Lyon-Caen, D. Hatzfeld and 9 other authors 2001. Crustal and upper mantle structure beneath the Corinth rift (Greece) from a teleseismic tomography study. *Journal of Geophysical Research* **105**, 28159–28171.

Trenhaile, A. S. 1987. *The geomorphology of rocky coasts*. Oxford: Oxford University Press.

Tzedakis, P. C., I. T. Lawson, M. R. Frogley, G. M. Hewitt, R. C. Preece 2002. Buffered tree population changes in a Quaternary refugium: evolutionary implications. *Science* **297**, 2044–2047.

Vita-Finzi, C. 1993. Evaluating Late Quaternary uplift in Greece and Cyprus. In *Magmatic processes and plate tectonics*, H. M. Pritchard, T. Alabaster, N. B. W. Harris, C. R. Neary (eds),

417–24. Special Publication 76, Geological Society, London.

Vita-Finzi, C. & G. C. P. King 1985. The seismicity, geomorphology and structural evolution of the Corinth area of Greece. *Royal Society of London, Philosophical Transactions A* **314**, 379–407.

Zelt, B. C., B. Taylor, M. Sachpazi, A. Hirn 2005. Crustal velocity and Moho structure beneath the Gulf of Corinth, Greece. *Geophysical Journal International* **162**, 257–68.

Index

159

INDEX